Laboratory Manual
For
Guinn and Brewer's

Essentials of General, Organic, and Biochemistry

First Edition

Sara Selfe
Edmonds Community College

W. H. Freeman and Company
New York

ISBN-13: 978-1-4292-2433-8
ISBN-10: 1-4292-2433-9

© 2010 by W. H. Freeman and Company

All right reserved.

Printed in the United States of America

First Printing

W. H. Freeman and Company
41 Madison Ave
New York, NY 10010
Houndmills, Basingstoke RG21 6XS England

www.whfreeman.com

Table of Contents

	Laboratory Safety	1
Experiment 1	Safety	5
Experiment 2	Material Safety Data Sheets (MSDS)	9
Experiment 3	Measurements and the Metric System	15
Experiment 4	Construction of a Density Column	25
Experiment 5	Solubility of Polar and Nonpolar Compounds	31
Experiment 6	Hot Packs and Cold Packs	39
Experiment 7	Diabetes and Dialysis	47
Experiment 8	Determination of the Calories in Food	55
Experiment 9	Stoichiometry and the Chemical Reaction	61
Experiment 10	Extraction of Caffeine from Beverages	71
Experiment 11	Using Vegetable Indicators to Determine pH	79
Experiment 12	Titration of Aspirin	85
Experiment 13	Vitamin C Content in Tablets and Beverages	91
Experiment 14	Synthesis of Soap	101
Experiment 15	Synthesis of Aspirin	109
Experiment 16	Synthesis of Acetaminophen	115
Experiment 17	Detection of Acetaminophen Poisoning	121
Experiment 18	Thin-Layer Chromatograpy of Analgesics	127
Experiment 19	Enzyme Specificity of Sucrose and Lactose Hydrolysis	133
Experiment 20	Hydrolysis of Sucrose	139
Experiment 21	Isolation of Casein from Milk	145
Experiment 22	Analysis of Wheat Germ Acid Phosphatase	151
Experiment 23	Radioactivity and Nuclear Medicine	161

LABORATORY SAFETY

Accidents will occur in the best-regulated families.
       Charles Dickens

While Dickens may have felt that his observation was true, the goal of laboratory safety is to prevent accidents in the "best-regulated laboratories." One of the most important facets of a chemistry course is teaching you how to work safely in the laboratory. With a good understanding of common safety rules, the laboratory is a very safe place to work. Your laboratory experience should be a safe one, if you heed the following precautions.

GENERAL SAFETY RULES

Safety goggles must be worn in the laboratory at all times.

Wearing goggles is one of the most important things you can do to protect yourself in a laboratory. The use of contact lenses in the laboratory is discouraged and, in some laboratories, prohibited. Check with your instructor.

Learn the location and operation of the safety equipment.

These items may include safety showers, emergency eyewashes, fire blankets, and fire extinguishers. If you are not familiar with the operation of a fire extinguisher, ask your instructor to explain it to you. Also, become familiar with all the exits from the laboratory. If a fire alarm goes off while you are in the laboratory, turn off any open flames and electric heaters, grab your valuables, and follow instructions given to you by your lab instructor.

Dress appropriately for the laboratory.

Bare feet, sandals, or other open-toed shoes are not a good idea in the laboratory. Cotton clothing (including denim) is particularly susceptible to being eaten by acid solutions. In general, the laboratory is not a good place to wear your favorite clothes. When working with flames, keep long hair tied back. Keep coats, backpacks, and other nonessential materials away from areas where people are working.

Do not eat, drink, or smoke in the laboratory.

To avoid contamination, do not bring consumable materials into the lab.

Do not leave a Bunsen burner or other heating apparatus unattended.

Turn off open flames if you must leave your area. Bunsen burner flames are often barely visible. Hair may catch on fire or a severe burn may result from leaning or reaching over an invisible flame.

Be aware of others in the laboratory.

When working in the laboratory you must be alert to those around you. Other students will be carrying chemicals to their workstation, so be careful when you walk through the laboratory. Also, if you are working with a flammable chemical check for open flames in the laboratory.

Working with Chemicals

Follow experimental procedures explicitly, checking and double-checking the identity of all reagents before you use them. There are potentially hazardous combinations of chemicals present in the laboratory. If you have an idea for further investigation, discuss it with your instructor and get authorization. Never attempt any unauthorized or unassigned experiments.

Clean up spills immediately.

The next person to come along has no way of knowing whether the clear liquid or white powder on the lab bench is innocuous or hazardous. Neutralize acid spills with sodium bicarbonate (baking soda) before cleaning them up. Mercury spills require special procedures; notify the lab assistant if there is a mercury spill.

Wash your hands frequently after handling chemicals and always before leaving the laboratory.

Never return unused reagents to their storage containers.

This is necessary to prevent contamination of the whole container. If you take more than you need dispose of the excess in the appropriate manner. Use the reagents sparingly—they are expensive and time consuming to prepare. When taking reagents, transfer the amount you need to a labeled clean beaker or other suitable container. Again, check and double-check the identity of all materials before using them.

Watch where you point things.

Do not point the open end of a test tube or other vessel containing a reaction mixture toward yourself or anyone else. If the procedure calls for you to observe the odor of the contents of a vessel, hold it upright 4 to 6 inches in front of you, gently fan some of the vapors toward your nose, and sniff gently. This is called wafting.

Always add acid to water slowly to dilute.

If you attempt to add water to concentrated acid, the heat of solution may vaporize the water and splash concentrated acid in your face. Sulfuric acid must be diluted particularly slowly because it releases a tremendous amount of heat upon dilution.

Dispose of reagents and other materials properly.

The proper disposal of reagents is essential to the health and safety of school faculty, staff, students, and the surrounding community. Reagents must be managed and discarded in the most responsible and environmentally sound method available. Only specified nonhazardous water-soluble materials can be rinsed down the drain. Waste containers for other materials will be provided. If you are unsure of how to dispose of a particular material, ask your instructor. Dispose of all broken glassware and other sharp objects into the appropriate container. Broken glass placed into the regular trash endangers the custodial personnel.

Wash chemicals off your skin immediately with copious amounts of water.

Use eyewash fountains if you get any chemicals in your eye (of course, this accident can be avoided by wearing goggles).

Hazard Identifications

Material Safety Data Sheets (MSDS) are provided by the manufacturer or vendor of a chemical. They contain information about the physical and chemical properties of the chemical and identify any hazards associated with the chemical. They also identify any special handling precautions and protective equipment needed when working with the chemical. You should be familiar with the MSDS before working with any chemical.

Read chemical labels carefully.

Chemicals are rated from 0 to 4 according to the hazard they impose, with 0 representing no hazard and 4 representing high hazard. An example of a hazard diamond label is shown below. Each chemical is rated for health, fire, and reactivity. Special warnings are reserved for the fourth diamond.

Hazard Diamond Colors

Fire = Red

Health = Blue

Reactivity = Yellow

Special Warning = White

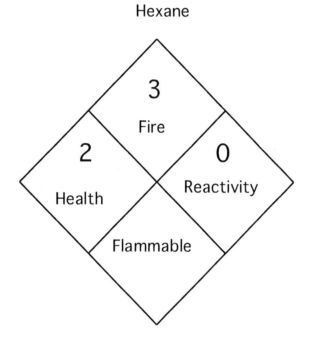

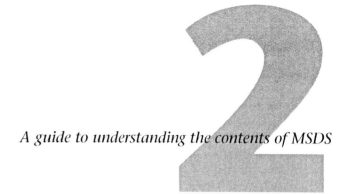

A guide to understanding the contents of MSDS

MATERIAL SAFETY DATA SHEETS (MSDS)

Material safety data sheets (MSDS) are provided by the manufacturer or vendor of a chemical. Employers are required to have MSDS available for review by employees. MSDS can be intimidating if you do not understand or are not familiar with the information they contain. In this lesson, we will look at MSDS and review the information it contains.

MSDS are divided into several sections. The first section generally contains the name of the compound, any possible synonyms, its chemical formula, and its CAS (**C**hemical **A**bstract **S**ystem) number. The CAS number is used as a method to unambiguously identify a compound. This system is needed because many compounds have one or more common names in addition to its formal chemical name. This lesson contains sample MSDS for sucrose. Look at the example and you will see that one synonym for sucrose is dextrose; but you probably know it as just plain "sugar."

The next MSDS section contains information about the physical properties of the chemical, such as color, density, boiling point, and solubility. Other sections contain information on reactivity, disposal, conditions to avoid, what to do in the case of a spill, and so on.

MSDS also identify any hazards associated with the chemical and list any special handling precautions and/or protective equipment needed when working with the chemical. Of particular importance is the health hazard, emergency, and first-aid data they contain. In the section that includes health hazard data, you will often find the term **LD_{50}**. This term is used as an attempt to quantify the degree of toxicity of a compound. The term LD_{50} indicates the dose of a chemical that is

The notation used is LD_{50} (ORAL, RAT): 5 mg/kg

lethal to 50% of the population. For example, strychnine sulfate, a violent poison, has an LD_{50} of 5 mg/kg when administered orally to rats. That is, if a population of rats is given strychnine sulfate in the amount of 5 mg/kg body weight, 50% of these rats will die.

Various abbreviations are used for both the species of animal tested and the mode of administration of the compound:

Animal tested		Mode of administration	
GPG	guinea pig	i.m. or mus	injected intramuscularly
MUS	mouse	i.p.	injected intraperitoneally
		i.v.	intravenously
		suc.	subcutaneously

Intraperitoneally means into the abdominal cavity.

In 2002, the LD_{50} Test was officially banned by the Organization for Economic Cooperation and Development (OECD). While new chemical products must be tested by an alternative methods, LD_{50} data will still be found in current MSDS.

Therefore, a notation of LD_{50} (i.p., GPG): 200 mg/kg, would mean that when a population of guinea pigs was given 200 mg of compound per kilogram of body weight via an injection intraperitoneally, 50% of these guinea pigs died. In addition to LD_{50}, you might see the term LC_{50}. LC stands for "lethal concentration." This can be used for concentrations of chemical in solution (e.g., water) or the concentration of a chemical in the air.

The trick in looking at LD_{50}'s is to be able to determine whether the compound is a severe poison or not harmful at all. You will see in the following MSDS for sucrose that LD_{50} is listed. However, we all know that sucrose or, as we call it, table sugar, is not generally considered toxic. Many of the warnings that are used by the manufacturer are stock phrases. For example, the sucrose MSDS suggest that you should induce vomiting if sucrose is ingested, something you probably *shouldn't* do when you use sugar at home. Even though this warning seems ridiculous, you cannot generally ignore the warnings given. The chemicals that are used in the lab are not manufactured to be consumed, so trace chemicals may exist that will be harmful—the warnings are real!

Another important section of MSDS is the listing of acute and chronic effects. Acute effects are those seen after a short exposure. As a student who may only be exposed to a particular chemical once, pay close attention to acute effects. Chronic effects are due to long-term exposure to a chemical. If you are an employee who works with chemicals, you will also need to be aware of the effects of exposure over time.

One of the goals of this lesson is to give you a sense of what these warnings and numbers mean and to help you gain a level of comfort when working with chemicals in the laboratory. One of the major purposes of taking a chemistry laboratory is for you to learn how to handle chemicals safely. Learning how to read MSDS is a step toward that goal.

To complete this lesson, you must use MSDS to determine specific information about sucrose (MSDS provided) and another compound of your choice. Several different compounds are given in the list in the sidebar as possibilities, or you may choose another compound you find interesting.

Choose one of the following:
acetic acid (found in vinegar)
acetylsalicylic acid (aspirin)
formaldehyde
hydrogen peroxide
iodine (used as an antiseptic)
isopropyl alcohol
menthol
methanol (wood alcohol)
naphthalene (mothballs)
sodium benzoate (preservative)
sodium bicarbonate
sodium chloride (table salt)

FINDING MSDS ON THE WEB

MSDS are easy to find on the Web using your search engine of choice. Using a browser, search for "MSDS." This search should give you a list of several databases. The easiest database to use is one with a searchable index.

MSDS for sucrose (abbreviated)

===
Ingredients/Identity Information
===

Ingredient: SUCROSE
Synonyms: CANE SUGAR, BEET SUGAR, DEXTROSE
Percent: 100.0
NIOSH (RTECS) Number: WN6500000
CAS Number: 57-50-1

===
Physical/Chemical Characteristics
===

Appearance and Odor: WHITE, ODORLESS CRYSTALS
Boiling Point: NOT GIVEN
Melting Point: DECOMPOSES
Vapor Pressure (MM Hg/70 F): NOT GIVEN
Vapor Density (Air=1): NOT GIVEN
Specific Gravity: 1.59
Decomposition Temperature: 220°F,104°C
Solubility in Water: FREELY SOLUBLE

===
Fire and Explosion Hazard Data
===

Flash Point: NONE
Lower Explosive Limit: NOT GIVEN
Upper Explosive Limit: NOT GIVEN
Extinguishing Media: DRY CHEMICAL, CARBON DIOXIDE, ALCOHOL FOAM
Special Fire Fighting Procedures: WEAR PROTECTIVE CLOTHING AND NIOSH-APPROVED SELF-CONTAINED BREATHING APPARATUS WITH FULL FACEPIECE OPERATED IN THE POSITIVE PRESSURE MODE.
Unusual Fire And Explosion Hazards: AS WITH ANY FINELY DIVIDED ORGANIC SOLID, DUST MAY BE EXPLOSIVE IF MIXED WITH AIR IN CRITICAL PROPORTIONS AND IN THE PRESENCE OF AN IGNITION SOURCE.

Reactivity Data

Stability: YES
Conditions to Avoid (Stability): EXTREME HEAT
Materials to Avoid: NITRIC ACID, SULFURIC ACID
Hazardous Decomp Products: CARBON MONOXIDE, CARBON DIOXIDE MAY BE FORMED.

Health Hazard Data

LD50-LC50 Mixture: LD50 (ORAL, RAT) IS 29700 MG/KG.
Route of Entry - Inhalation: YES   Route Of Entry - Skin: YES
Route of Entry - Ingestion: YES
Health Hazard Acute and Chronic: EYE CONTACT MAY CAUSE IRRITATION SKIN CONTACT MAY CAUSE IRRITATION. INGESTION MAY CAUSE GASTROINTESTINAL DISCOMFORT. INHALATION MAY CAUSE IRRITATION TO RESPIRATORY TRACT.
Carcinogenicity : NO
Signs/Symptoms of Overexp: EYE IRRITATION, SKIN IRRITATION, GI TRACT IRRITATION, RESPIRATORY TRACT IRRITATION.
Emergency/First Aid Proc: EYES: FLUSH WITH WATER FOR 15 MINUTES WHILE HOLDING EYELIDS OPEN. GET MEDICAL ATTENTION. SKIN: REMOVE CONTAMINATED CLOTHING. WASH WITH SOAP AND WATER. IF IRRITATION PERSISTS, GET MEDICAL ATTENTION. INHALATION: REMOVE TO FRESH AIR. RESTORE BREATHING. GET MEDICAL ATTENTION. INGESTION: GIVE WATER AND INDUCE VOMITING AS DIRECTED BY MEDICAL PERSONNEL. IMMEDIATELY CALL A PHYSICIAN OR POISON CONTROL.

Precautions for Safe Handling and Use

Steps If Matl Released/Spill: VENTILATE AREA OF SPILL. CLEAN-UP PERSONNEL SHOULD WEAR PROPER PROTECTIVE EQUIPMENT. AVOID CREATING DUST. SWEEP OR SCOOP UP AND CONTAINERIZE FOR DISPOSAL.
Waste Disposal Method: WHATEVER CANNOT BE SAVED FOR RECOVERY MAY BE BURNED IN AN APPROVED INCINERATOR OR DISPOSED IN AN APPROVED WASTE FACILITY. ENSURE COMPLIANCE WITH LOCAL, STATE, AND FEDERAL REGULATIONS.
Precautions-Handling/Storing: STORE TIGHTLY CLOSED IN A COOL, DRY, WELL-VENTILATED AREA. SUITABLE FOR STORAGE IN ANY GENERAL CHEMICAL STORAGE AREA. DO NOT BREATHE DUST.

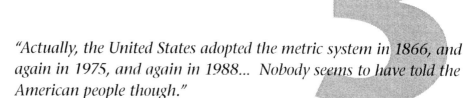

"Actually, the United States adopted the metric system in 1866, and again in 1975, and again in 1988... Nobody seems to have told the American people though."

The Famous Druid on Geek Culture Forum

MEASUREMENTS and the METRIC SYSTEM

As students move into the sciences (biology, chemistry, etc.) they need to become familiar with the metric system. In the United States the metric system is the not the primary mode of measurement in commerce. If you grew up in the United States, you are more familiar with pounds, yards, and gallons than you are with grams, meters, and liters. But science and medicine use the metric system, and this lab is design to familiarize you with metric units.

METRIC SYSTEM

The primary units of the metric system are:

- The **meter** for distance or length
- The **gram** for mass
- The **liter** for volume

The metric system is a base 10 system. You are familiar with the base 10 system with the dollar. There is a penny (one cent) that is 1/100 (0.01 or 10^{-2}) of a dollar and a dime that is 1/10 (0.1 or 10^{-1}) of a dollar. In the metric system, instead of having a different word for each unit (penny and dime), prefixes are used to tell you what multiple of ten you need to either multiply by or divide into the base unit. For example, centi, abbreviated "c," is the prefix for 1/100 of a base unit. Therefore, if the dollar is the base unit, then a penny is a centidollar (probably why they called it a cent). In the metric system, if you were measuring a length of one centimeter (cm), that would be 1/100 of a meter. It really is much easier than having to remember inches, feet, yards, miles, etc.; you just need to know a meter and then you can have any multiple of ten (factor) of a meter by just adding a prefix.

TABLE of COMMON METRIC PREFIXES

Prefix	Symbol	Factor	Factor in Scientific Notation
kilo	k	1 000	10^3
deci	d	0.1	10^{-1}
centi	c	0.01	10^{-2}
milli	m	0.001	10^{-3}
micro	µ or mc	0.000001	10^{-6}

CONVERSION FACTORS

Conversion factors are used to convert one unit into another. If you want to convert pennies to dollars, you would use the conversion factor 1 dollar = 100 pennies. If you have 500 pennies and you want to convert to dollars, you would take 500 pennies × 1 dollar/100 pennies = 5.00 dollars. Conversion factors are great because you can use the same conversion factor to change into dollar or cents. If I have 5.00 dollars and want to convert to pennies, all I need to do is take 5.00 dollars × 100 pennies/1 dollar = 500 pennies. Once you establish the appropriate equality, in this case 1 dollar = 100 pennies, you multiply the supplied unit (for example, dollar) by the requested unit/supplied unit. In this last example, the requested unit is penny and the supplied unit is dollar.

Now for metrics. If I have 475 cm, how many meters do I have? The conversion factor is 1 meter = 100 centimeters. Looking at the table above, the factor for centi is 0.01, therefore, 1 centimeter = 0.01 m. If I want to convert to meters, I take 475 cm × 0.01m/cm = 4.75 m. The same converson factor can be used to convert the 4.75 m into centimeters by multiplying by cm/0.01m (the requested unit, cm/ the supplied unit, m). Therefore, 4.75 m × cm/0.01 m = 475 cm.

SIGNIFICANT FIGURES and CHOOSING APPROPRIATE GLASSWARE

Whenever you measure something, there is some guesswork involved. Let's say you are measuring the line below using the "ruler" underneath it. The length of the line is greater than 6 units, but not quite 7. So you make a guess: I might say 6.5 units, but you might say 6.6 units. Scientists like to say this last digit is uncertain, probably because it sounds a little better than saying "It's a guess." This measurement has two significant figures, one that we know for sure and another digit that is uncertain Now if I were to say it was 6.654 units, that would just be unrealistic — there is no way I can guess that well. If I say it is just 6, that isn't really close either, so you measure as close as you can, with the last digit carrying some uncertainty.

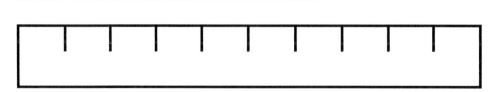

The number of significant figures you can measure depends on the ruler, glassware, balance, etc., you use. The more precise the instrument, the more significant figures you will have. Now if you were trying to measure something in the laboratory, you would pick a better ruler than the one shown on page 16, one with a finer scale. For instance, you want to measure out 26 mL of a liquid and you have two choices, a 50-mL graduated cylinder and a 50-mL beaker.

Below are examples of the scales of each of these. A 50-mL beaker on the right only has a scale that measures 10, 20, 30, and 40 mL whereas a 50-mL graduated cylinder has a scale that measures every mL.

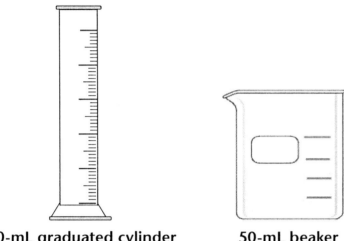

50-mL graduated cylinder   **50-mL beaker**

If you use the 50-mL beaker, you can get a fairly good 20 mL, but you are really guessing to get 26 mL. The graduated cylinder has a mark for every mL, therefore, you actually have a mark for 26 mL — that's the one I would choose. In the case of reading this graduated cylinder, you can still read an uncertain figure. If the height of liquid you are measuring falls between 26 and 27 mL, you can estimate a digit for a tenth of a mL. That is, if the height falls right to the 26 mL mark, then you would report 26.0 mL. If the height was about halfway between 26 and 27 mL, you might report it as 26.5 mL — you make your best estimate of the volume. So when using glassware, you report the digits you know from the scale given and estimate any amount that falls between two marks on the scale.

To determine the number of significant figures for the beaker above, you first see that you have marks for 10, 20, 30, and 40 mL, and each of these has one significant figure (the zero is only there to tell us it is 40 and not 4). You next estimate the number between the lines on the scale and report two significant figures (like the estimation of 26 mL in the example above). The graduated cylinder has a scale that measures each mL: 1, 2, 3, ...49, 50. Again, if you were measuring 26 mL, you would report two significant figures that you know for sure and a third significant figure that is any distance between the two marks on the scale.

The significant figures you have in your data all come from the measurements you make. Choosing the right piece of glassware for measurement will make your experiment more accurate. You would not use a 1-foot ruler to measure the length of the lab. It would not only be inaccurate (every time you move the ruler there is a chance of error), it would just take too much time. Choosing the appropriate tool when making any measurement is part of good lab technique.

In this lab, use the appropriate number of significant figures for each measurement. If you are unsure how to correctly read the glassware, ask your instructor.

MATERIALS

Equipment
small ruler (with mm)
meter stick
50-mL graduated cylinder
100-mL beaker
irregularly shaped object
thermometer

Chemicals
sugar cubes
cream packets
ice

PROCEDURE

For each of the sections, write your data in the Data and Results section that follows.

Part A: Measuring Length

1. Using a 1-meter stick, measure your height. This can most easily be done by standing against the paper provided on the wall in the lab (DO NOT DRAW ON THE WALLS!) and asking a colleague to mark your height on the wall. Measure and record your height in centimeters.

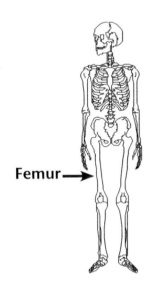
Femur→

2. In forensics and anthropology, the length of the femur (the leg bone that runs from the hip to the knee) is used to estimate height when a full skeleton is not available. With the help of your partner, measure the length of your femur. Record the height and femur length for at least three people in the data table.

Part B: Measuring Volume

1. Take a sugar cube and measure each side in mm. The volume of the cube can be determined by multiplying the length × height × width. Keep the sugar cube; you will use it in Part C.

2. Measure the volume of "cream" in the packet provided in units of milliliters, using your 50-mL graduated cylinder. [See below for instructions on how to correctly read the volume of a liquid.] Record this value on your data sheet. Check the packet for a volume given by the manufacturer and also record this value on your data sheet. Wash your graduated cylinder with a brush (and soap, if necessary).

Reading Volume of a Liquid:

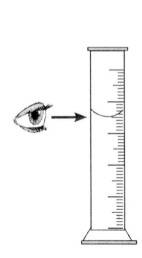

To read the volume of a liquid in a graduated cylinder, it is necessary to have your eyes level with the liquid level. In the case of water and some other liquids, you will notice that the surface of the liquid is not a straight line, but a curve. This curvature is called a meniscus and is due to the surface tension of the liquid. When determining the volume when a meniscus is present, you read the volume at the bottom of the curve

Part C: Measuring Mass

If you do not know how to use the lab balance, ask your instructor for assistance.

1. Choose any piece of glassware from your drawer and determine its mass.
2. Measure the mass of the sugar cube on the lab balance.

3. Look at the package of sugar cubes and find the mass the manufacturer states for each sugar cube.

Part D: Measuring Density

50.9g , 120.2 , 91.9

1. Measure the mass of a clean 100-mL beaker. Measure 40 mL of water using a clean, dry 50-mL graduated cylinder. Record this volume of water to the appropriate number of significant figures. Add this water to the previously weighed 100-mL beaker and record the mass. Keep this water for Part E.

2. Measure the mass and volume of an irregularly shaped solid object.

    a. Measure the mass of the object first.

    b. After you have measured the mass, determine the volume by water displacement. If you submerge the object in water, it will displace its volume of water. Add water to the 50-mL graduated cylinder until it is half-full. Record the volume. Drop the object (gently, no splash) into the water and record the new volume. The difference between the two volume measurements is the volume of the object.

A Roman mathematician, Archimedes, born in 287 BCE, realized the principle of using water displacement to determine volume when he was in his bathtub.

Part E: Measuring Temperature

1. The beaker of water in the previous section has probably come to room temperature by now. Measure the temperature of the water in Celsius.

2. Add about 20 mL of ice to the water and wait 5–10 minutes. Measure the temperature of the ice water in Celsius.

Exploration of the physical properties of density and solubility

CONSTRUCTION of a DENSITY COLUMN

You have probably heard the terms LDL and HDL, low-density lipoproteins and high-density lipoproteins. What about bone density? As we age we lose bone density, which is why calcium, vitamin D, and weight-bearing exercise (walking, for example) are recommended to counter this loss. Density is a physical property that is used to characterize substances, such as lipoproteins, bone, water, metals, etc. In this experiment, you will be use the **physical property** of density to construct a column composed of four distinct liquid layers. In the column, the position of the layers is due to the difference in density.

Physical properties are the properties of a substance that we can observe or measure that do not change the substance.

Density is defined as the mass of a material per given unit of volume. It is probably a familiar concept; we know a metal weight will sink in water because it weighs more than the same volume of water. To determine the density of an object, you need two pieces of information: the volume and the mass of that volume. The density of a liquid is determined by using the expression:

$$\text{Density} = \frac{\text{mass in grams}}{\text{volume in mL}}$$

The density of a liquid is given in grams per milliliter (g/mL) because the milliliter is the unit commonly used to measure liquids. A different unit is used for solids. The volume of a solid, if it is a cube, can be determined by multiplying the height (in centimeters, cm), the width (in cm), and the length (in cm) to get cm^3. Therefore, the mass of a solid is given in grams per cm^3.

In the construction of the column, you will also use another physical property—solubility. If you have done any cooking, you realize that oil is not soluble in water. You can mix oil and water by shaking, but in time, they will separate. By contrast, ethanol is soluble in water; they mix completely and do not separate upon standing. The old saying "like dissolves like" summarizes this property. Compounds that are similar in their interactions with each other in most cases will be soluble. Thus polar materials are soluble with other polar materials and nonpolar materials are soluble with other nonpolar materials. However, polar and nonpolar materials are generally not soluble with each other.

The four liquids you will be placing in your density column are water, glycerol, saturated potassium bromide solution, and hexane—each with different densities and polarities. The potassium bromide solution in water is strictly polar and hexane is strictly nonpolar, but the glycerol has a molecular structure that results in a polarity that is intermediate between polar and nonpolar. In this experiment, you will determine the different densities of these liquids in order to create a density column. You will then drop different solids into your density column. Blocks will settle below a layer of a lesser density and on top of a layer of greater density. You will then use this information to determine the approximate densities of different blocks of solids.

MATERIALS

Equipment
10-mL graduated cylinder(s)

50-mL graduated cylinder

100-mL beaker

4 small containers

spatula

blocks made from different materials

Chemicals
glycerol, $C_3H_8O_3$

saturated potassium bromide solution, KBr in H_2O

hexane, C_6H_{14}

water, H_2O

CAUTION
Hexane is flammable, so keep it away from open flames. Dispense hexane in the hood. Hexane should not be disposed of down the drain.

PROCEDURE

Part A: Determination of the Density of the Liquids

1. Weigh a 10-mL graduated cylinder. Record the mass in the Data section.
2. Add exactly 10 mL of one of the liquids to the graduated cylinder.
3. Weigh the graduated cylinder plus liquid. Record the mass in the Data section.
4. In the Results section, calculate the density of the liquid by using the following equation:

$$\text{Density} = \frac{(\text{mass of graduated cylinder + liquid}) - (\text{mass of graduated cylinder empty})}{\text{mL}}$$

5. Put the first liquid aside in a container. Repeat steps 1 through 4 for each liquid. If you must use the same graduated cylinder, be sure to rinse out all traces of the previous liquid before adding a new liquid. After weighing, place each liquid in a separate container until ready to make the density column.

Part B: Construction of the Density Column

After you have calculated the densities of the individual solutions, you can begin the construction of the density column. Verify the order before beginning.

1. Carefully pour the 10 mL of the densest liquid into a 50-mL graduated cylinder.

2. Carefully pour the 10 mL of the next densest liquid into the 50-mL graduated cylinder. Pour gently to minimize mixing of layers.
3. Carefully follow this with the next densest liquid.
4. Finally, add the least dense liquid, being careful not to allow the liquids to mix. If you have poured the liquids carefully, you should have a four-layer column at this point. Record your observations.
5. Determine the approximate density of two different materials supplied in the laboratory by gently dropping them into the liquid column. You may need to apply a slight pressure (poke) to the materials to break any surface tension that exists. Record between which two layers the material rests.
6. Mix the layers in the column by pouring the contents of the column back and forth five times between the graduated cylinder and a 100-mL beaker. Record your observations.

Name: _____

DATA

	Mass of graduated cylinder + liquid	Mass of graduated cylinder	Mass of liquid
glycerol			
hexane			
saturated KCl			
water			

RESULTS

1. Calculate the densities of the four liquids. Show a sample calculation below. $\text{Density} = \dfrac{\text{mass of liquid}}{\text{mL}}$

2. List the densities of the four liquids.

   glycerol _____

   hexane _____

   saturated KCl _____

   water _____

3. List the liquids in the order in which you will add them to the column. The solutions are added to the column by density: The most dense is added first, and the least dense is added last. Verify that you will be adding the most dense solution first.

   Solution added to column first (most dense) _____

   Solution added to column second _____

Solution added to column third _____

Solution added to column last (least dense) _____

4. Draw a picture of your density column. Label the layers with the names of the liquids.

5. Determine the approximate densities of two solids. Report the densities as a range. A solid that settles between layers one and two is more dense than layer one and less dense than layer two. Be sure to note which solids you used.

6. What happened when you mixed the layers in the column? Include in this description whether you can determine which liquids are soluble in each other.

7. Bone density varies with the type of bone and the age of the person. An average for bone density in a healthy adult is about 1.3 g/cm^3. Where would a fragment of bone of this density be in your density column? Elderly women can lose up to 50% of their bone density. Where would a fragment of their bone be found?

Like dissolves like.

SOLUBILITY of POLAR and NONPOLAR COMPOUNDS

One of the most important physical properties of a chemical compound is its solubility. When one compound is soluble in another, they form a solution. A solution is a homogeneous mixture that generally contains a large amount of one component (solvent) into which a small amount of another component (solute) is dissolved.

You have probably observed that some pairs of liquids, such as oil and water, will not mix, whereas others, such as alcohol and water, will mix readily. Two compounds that will dissolve in one another are said to be **soluble** and form a solution as described above. Those that will not dissolve are referred to as **insoluble**. The common axiom used to describe this is "like dissolves like."

Solubility of liquids (or lack thereof) can be explained by a simplified discussion of polarity. In chemistry, we generally divide compounds into two large groups: polar and nonpolar. In biological systems, we talk of chemicals as being water-soluble (polar) or fat-soluble (nonpolar). Polarity is based on the distribution of electrical charge on the molecule of the compound. Molecules that have specific areas of positive and negative charge are polar. Water and alcohol are examples of polar molecules. In the molecules of a nonpolar material, the charge is evenly distributed; therefore, there are no positive and negative regions. Because the magnitude of charge distribution can vary, a range of polarities can exist. It is important to remember that the total charge on any **molecule** equals zero. In polar molecules, however, the positive and negative charges are each relegated to distinct

Miscibility is a special case of solubility. When two liquids are miscible, they will mix in all proportions, or are said to be infinitely soluble.

A water molecule is polar.

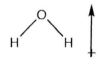

regions. The extreme of distributed charges is found in ionic compounds, where you have distinctly separate ions with positive and negative charges.

Although both sugar and table salt readily dissolve in water, there is a significant difference in the two solutions. The salt solution is able to conduct electricity, whereas the sugar solution cannot. This ability to conduct electricity when dissolved in water is characteristic of soluble ionic compounds. When ionic compounds dissolve in water, they form a solution that contains the separate ions. These solutions conduct electricity, and the dissolved compounds are termed **electrolytes**. Some covalent compounds form ions in solution and are also electrolytes. You can test for the presence of electrolytes by using a conductivity tester.

> Electrolytes are substances that produce ions in solution.

Electrolytes are significant in biological systems because they are important in maintaining cell performance. Particular ions are required for specific functions in the body. Sodium and potassium ions are the major cations present in extracellular and intracellular fluid, respectively. There are many cases where you need to supply electrolytes to your systems. When there is a loss of fluids and/or electrolytes, a solution may be administered to replenish them. Drinks such as Powerade™ and Gatorade™ provide electrolytes and carbohydrates during exercise. Babies and young children can become dehydrated and lose electrolytes when ill with diarrhea and vomiting. Doctors often prescribe solutions, such as Pedialyte™, to supply fluids and electrolytes.

Solubility is also affected by temperature. Most solids have an increased solubility at higher temperatures. Temperature has a large effect on the solubility of gases in water. As the temperature increases, gas is less soluble in water. You notice this effect in everyday life when you leave a soft drink open at room temperature. As the temperature of the drink increases, the carbon dioxide gas (the carbonation) escapes and your pop loses it "fizz."

MATERIALS

Equipment
small test tubes
test-tube rack
spatula
wood splints
10-mL graduated cylinder
dropper
beakers
hot plate or heat source
conductivity tester
thermometer
marking pencil
100-mL graduate cylinder
100-mL beakers
scissors

Chemicals
sucrose, $C_{12}H_{22}O_{12}$
iodine, I_2
anhydrous ammonium chloride, NH_4Cl
anhydrous calcium chloride, $CaCl_2$
sodium chloride, $NaCl$
potassium permanganate, $KMnO_4$
cholesterol, $C_{27}H_{46}O$
ascorbic acid (vitamin C), $C_6H_8O_6$
ethanol, C_2H_5OH
vegetable oil
vitamin E tablets
hexane, C_6H_{14}
bromthymol blue
Perrier or other carbonated bottled water

> **CAUTION**
> Iodine can irritate the skin and eyes; handle it with particular caution.

> **WASTE DISPOSAL**
> Both hexane and ethanol are flammable. These liquids cannot be poured down the drain. Place excess or waste liquids into the appropriate waste containers.

PROCEDURE

Part A: Solubility in Polar and Nonpolar Solvents

Don't forget to put on your goggles!

1. Label 12 test tubes 1 through 12.

2. Using a graduated cylinder, measure and pour 3 mL of water (distilled, if available) into a test tube. Mark the height of this volume on test tube #1 with a marking pencil. Place a mark at the same height on the eleven other labeled test tubes. Add 3 mL of water to each test tube by filling up to the mark.

3. Test the water solubility of each of the twelve solutes by adding a different solute to each test tube. The twelve solutes are: sucrose, iodine, ammonium chloride, calcium chloride, sodium chloride, potassium permanganate, cholesterol, vitamin C, ethanol, vitamine E, hexane, and vegetable oil. For liquid solutes, add approximately 1 mL. For the vitamin E, clip off the end of the tablet and squeeze the vitamin E into the test tube. For solid solutes, transfer a small amount (about the size of small pea) of each solid solute, using a metal spatula or wood splint. Because iodine reacts with some metals, you must use a wood splint for iodine; discard the splint after use.

4. Gently mix each test tube's contents using your finger to firmly tap on the side of the test tube.

5. Make judgments about the solubility of each solute: soluble, slightly soluble, insoluble. Also, note any changes in color, temperature, and so forth. For liquids note whether the solute is less dense (is on top) or more dense than water.

When some compounds are dissolved in water, the resultant solution can be quite warm or cold. This is the basis for some hot and cold packs.

6. Test the conductivity of pure water. Insert the probes of the conductivity tester into the water and note any illumination of the light emitting diode (LED). Distilled water should not conduct electricity. If you used tap water, note the brightness of the LED. You will be able to make comparative measurements of conductivity. Anything "brighter" than pure water means that there are some electrolytes present.

7. Test the conductivity of your solutions. If the probes do not reach down to the solution, transfer your solution to a small beaker for the test. Remember to rinse the probes after each test.

8. Discard all test tube contents into labeled waste containers.

9. Wash and thoroughly dry the test tubes.

10. Repeat steps 2 through 5, using the nonpolar solvent hexane instead of water.

Part B: Temperature Effect on the Solubility of Gases

Many bottled waters and soft drinks are carbonated because they contain dissolved carbon dioxide gas. Carbon dioxide forms an acidic solution when dissolved in water as a result of the formation of carbonic acid. This is the same equilibrium seen in the bloodstream when carbon dioxide is present. Excess carbon dioxide in the bloodstream can lead to respiratory acidosis, the presence of excess acid (H^+).

$$H_2O + CO_2 \leftrightarrows H_2CO_3 \leftrightarrows H^+ + HCO_3^-$$

Carbonic acid

If the solubility of carbon dioxide gas changes with increasing temperature, you should see a change in the acidity of a carbonated solution upon boiling. For this part of the experiment, you will use an indicator (bromthymol blue) that is yellow at low pH (acid solution) and blue at high pH (neutral or basic solution).

The bottle of carbonated water must be freshly opened so that it has not gone "flat," that is, lost all its carbon dioxide.

1. Open a bottle of carbonated water and dispense 30 mL into each of two beakers.
2. Add three drops of bromthymol blue indicator to each beaker. Note the color of the solution.
3. Heat one of the beakers on a hot plate for 5–10 minutes (or to boiling). Allow the solution to cool. Note the color of the solution after heating.

Part C: Temperature Effect on Solubility of Solids

In contrast to gases, the solubility of most solids increases with temperature. You know from experience that sugar dissolves more readily in hot tea than in iced tea. We will test this in the following part of the experiment.

1. Fill two 100-mL beakers with water. Heat one beaker to 70°–80°C.
2. Add several (4–5) crystals of potassium permanganate ($KMnO_4$) to each beaker.
3. Describe the intensity of the color in each beaker over time: immediately after addition, 1 minute, 5 minutes, up to 15 minutes.

Heat, temperature, and specific heat

HOT PACKS and COLD PACKS

Depending on the circumstances, a doctor or nurse may prescribe the use of hot or cold treatments to aid in the relief of pain. The application of heat will increase circulation and help reduce inflammation. By contrast, cold is used to reduce swelling and decrease pain. Commercially available hot packs and cold packs utilize the heat given off or absorbed when compounds dissolve or crystallize. In this experiment, you will observe the temperature changes that occur when calcium chloride and ammonium chloride are dissolved in water. From these data, you will be able to calculate the heat given off or absorbed during the dissolution.

Heat is not the same as temperature. **Temperature** does not depend on the quantity of the material; it is simply a measure of how hot or cold a sample is. **Heat,** on the other hand, does depend on the quantity of the material and is a measure of how much energy a sample contains. For example, compare the temperature and the heat content of a cup of freshly brewed coffee and an urn of freshly brewed coffee. The coffee in each case will be approximately the same temperature, but an urn full of coffee contains more heat than a cup of coffee simply because there is more coffee.

Heat can be measured in °C (Celsius), °F (Fahrenheit), or K (Kelvin).

Heat content is dependent not only on the quantity of a substance, but also on the identity of the substance. Each substance has its own specific heat. The specific heat of water is 1.0 calorie/g°C. This statement means it takes 1.0 calorie of energy to raise the temperature of 1.0 g of water by 1 degree Celsius. Ethyl alcohol has a specific heat of only 0.59 calories/g°C. Much less energy is needed to raise the temperature of 1 g of ethyl alcohol by 1°C.

Specific heat is defined as the amount of heat needed to raise the temperature of 1 g of the substance by 1°C.

To calculate the heat given off or absorbed during a dissolution reaction, there is a third aspect to consider (besides the mass and the specific heat of the material): temperature change (ΔT). The calculation is expressed as

Heat given off or absorbed = (specific heat)(mass)(temperature change)

The particular reactions in this experiment are those that take place in commercially available hot and cold packs. The packs consist of an outer pouch containing the solid and an inner pouch containing water. When the inner pouch is broken, the dissolution takes place with an accompanying release or absorption of heat. Some hot packs are based on the heat given off when a saturated solution crystallizes.

A process that gives off heat is called **exothermic,** and a process that absorbs heat is called **endothermic.**

MATERIALS

Equipment
thermometer
stirring rod
spatula
4 Styrofoam cups
100-mL graduated cylinder
stopwatch

Chemicals
calcium hydroxide, $CaCl_2$
ammonium chloride, NH_4Cl

WASTE DISPOSAL
The calcium chloride and ammonium chloride are nonhazardous and can be rinsed down the drain with copious amounts of water.

SPECIAL NOTES

This experiment is quantitative, so be careful when weighing. You do not need to weigh out exactly 20 g, but you do need to record the mass of the compound after weighing.

A **calorimeter** is an apparatus used to measure heat.

PROCEDURE

1. Using two Styrofoam cups, construct a simple calorimeter by nesting the two cups, one inside the other.
2. Add 100 mL of water to the calorimeter. Allow the water to stand for several minutes to reach a stable temperature. Record the temperature; this is the initial temperature (T_i).
3. Weigh out approximately 20 g of $CaCl_2$. Be sure to record the exact mass.
4. While your partner holds the cup and the thermometer steady, add all the $CaCl_2$ to the calorimeter and stir rapidly with a stirring rod. Be careful not to hit the thermometer while stirring.

CAUTION
Do not use the thermometer as a stirring rod.

5. After mixing, time-temperature points should be recorded in the data table. One partner should read the temperature while the other reads the time and keeps the record. It is best to take temperature readings at frequent intervals, every minute or less, while the temperature is changing dramatically. Take readings at longer intervals, every 2 to 3 minutes, as the temperature change starts to slow.

You will make a plot on graph paper to find the maximum temperature. The maximum temperature reached is the final temperature (T_f).

6. After recording your data, wash the contents of the cup down the drain with lots of water. Rinse off the thermometer and stirring rod.

7. Repeat the above procedure, using approximately 20 g ammonium chloride (NH_4Cl). Be sure to record the exact mass. The dissolution of ammonium chloride is endothermic, so in this case, the minimum temperature reached is final temperature (T_f).

The American Diabetes Association reports that 20.8 million children and adults in the United States have diabetes.

DIABETES and DIALYSIS

Kidneys process our blood on a daily basis. Their function is to remove excess water and waste products from our blood. In performing this function, kidneys use the process of dialysis. Dialysis is the movement of solutes (small molecules and ions) across a semipermeable membrane. A semipermeable membrane is a structure that allows the passage of certain solutes and molecules while preventing the passage of other molecules and colloidal particles across the membrane. For example, the kidneys remove small molecules such as urea, glucose, and excess salts but retain large molecules such as proteins. Excess water is also removed by the kidneys. Since this is the movement of the solvent, it is referred to as osmosis. Both of these processes are driven by a concentration gradient (a difference in concentration) —solvents move toward areas of high concentration and solutes move toward areas of low concentration. In both cases, the net effect is to try to establish solutions with the same concentration.

Useful molecules such as glucose that are removed during the above process are reabsorbed by the kidneys *except* when there is excess glucose present, as there is in the case of diabetes. One of the first indications of diabetes that a doctor may note is the presence of glucose in a patient's urine. The blood sugar level of a diabetic can be 25% higher than that of a person without diabetes. This glucose must be removed or organs such as kidneys and eyes can be damaged.

Two of the symptoms of diabetes, excessive thirst and frequent urination, are linked to the high concentration of glucose in the blood. The removal of excess glucose from blood by the kidneys increases the concentration of glucose in urine. This increase in concentration causes water to move into the urine to dilute the solute

The component that has the lower concentration in a solution is the **solute**, and the one that has the large concentration is the **solvent**.

Dialysis is the transport of solutes (generally small molecules and ions) across a semipermeable membrane.

Osmosis is a process where solvent moves across a semipermeable membrane from an area of low concentration to high concentration.

(glucose), which leads to frequent urination in diabetics. This loss of water from the body also causes the increased thirst.

In this experiment, you will use dialysis tubing to look at the process of dialysis and osmosis. Dialysis tubing has small pores that will allow the passage of small molecules while retaining larger molecules—a semipermeable membrane. There are different types of dialysis tubing based on the size of the pores and the type of material used to make the tubing. In general, dialysis tubing is categorized by its pore size. So one might purchase tubing with a "1000 MW (molecular weight) cutoff." This would mean that only molecules with a MW less than 1000 g/mol would pass through the pores. So if we were to dialyze a solution that contained sucrose (MW = 342 g/mol) and starch (MW > 150,000 g/mol) in tubing with a 1000 MW cutoff, the sucrose would pass through the pores while the starch would not.

The solution you are going to dialyze is honey. What are the components of honey? Honey is mostly invert sugar, a mixture of fructose and glucose. There are other things in honey, such as B vitamins (no pun intended), vitamin C, minerals, water, and proteins (for example, invertase, which breaks down sucrose into invert sugar). Some of these components can move through the pores and others cannot. In the pre-lab you will first try to determine what you expect to happen, then you will do the experiment and see if you are right!

MATERIALS

Equipment
dialysis tubine

watch glass

4 250-mL beakers

test tubes

funnel

eyedropper or pipet

10-mL graduated cylinder

Chemicals
honey

50% honey (aqueous)

Benedict's reagent

PROCEDURE
Part A: Dialysis/Osmosis

CAUTION

It is important to use the funnel so that no honey gets on the outside of the dialysis bag. Any honey on the outside will contaminate the water in the beaker.

1. Place two 7-inch pieces of dialysis tubing in water for 10 minutes to hydrate.
2. After 10 minutes, remove one piece of dialysis tubing from the water and gently tie a knot about 1 inch from one end.
3. Take the other end of the tubing and rub it between two fingers to open the tubing (like opening plastic bags in the grocery store). Insert the funnel into the tubing and add about 5 mL (it is about an inch) of honey to the tubing and then tie a knot about 1 inch from the top. You must be careful not to get honey on the exterior of the bag or it will give a false positive test in part B.
4. Weigh a watch glass and record the mass in the Data and Results section.
5. Place the dialysis bag with the honey on the watch glass and weigh. Record the total mass in the Data and Results section.

6. Place the dialysis bag in a 250-mL beaker with sufficient water to cover the dialysis bag (about 100 mL). Record the start time on the data sheet.
7. Dialyze the honey for 20 minutes. Record the end time on the data sheet.
8. While the honey is dialyzing, repeat from step 2 above for the 50% aq. honey solution.
9. After 20 minutes, remove the dialysis bag, record the time, and gently wipe off the water from the surface of the bag.
10. Re-weigh the dialysis bag (don't forget the watch glass) to determine if there was any change in mass.

Part B: Testing for Presence of Glucose and/or Fructose

Both glucose and fructose are reducing sugars. A test for the presence of reducing sugars is a Benedict's test, where a solution containing copper is added to the sugars and then heated. The blue copper solution changes to colors ranging from mustard yellow to brick-red if a reducing sugar is present. If fructose or glucose dialyzed out of the bag, you should see their presence in the water in the beaker. You will be doing four tests: 1) water outside the dialysis bag of honey, 2) water outside the dialysis bag of 50% aq. honey, 3) water (negative control), and 4) 50% honey (positive control). Of course, you must be careful not to accidently contaminate the samples with any honey or you might get a positive test due to the contaminant.

1. Place four test tubes in a test tube rack and label them 1 through 4 to correspond to the solutions described above.
2. In test tube 1, place 1 mL of the the water from outside the dialysis bag with honey.
3. In test tube 2, place 1 mL of the the water from outside the dialysis bag with 50% aq. honey.
4. In test tubes 3 and 4, place 1 mL of the water. To test tube 4, add one drop of the 50% aq. honey solution.
5. Place 1 mL of Benedict's solution in each of the four test tubes.
6. Place all four test tubes in a beaker of gently boiling water. Any change in color should occur within 5–10 minutes
7. Record any change in color on the Data and Results sheet.

The journey of a thousand pounds begins with a single burger.
Chris O'Brien

DETERMINATION of the CALORIES in FOOD

Fat, trans fat, low fat, fat free. High fructose corn syrup, sugar, sugar substitutes. You see the labels everywhere. Whether you are watching your weight or not we are all aware of calories. For many of us it is one of the things on the food labels we look at first. Monitoring your calorie intake is one of the important steps to weight maintenance. Calories come from three sources: fats, carbohydrates, and proteins.

Fats are high-energy molecules. One gram of fat can release 9000 calories (9 food Calories) of energy, which is more than double the energy released from an equal mass of carbohydrate or protein. It is not surprising that the body produces fat to store excess food energy efficiently and that it is difficult to "burn off" excess fat. Gram for gram, you must run more than twice as far or exercise twice as long to "burn up" fat as you do to burn off carbohydrates. Our body stores fat for a "rainy day," just in case we need it due to famine. We also store carbohydrates for quick fuel, but fat is our long-term storage. You can think of carbohydrates as the food in the refrigerator to be used fairly quickly and fat as either freezer food or canned goods — for when the refrigerator is empty. Protein is another source of energy, but it is not generally used as an energy source for the human body (except under starvation conditions).

In this experiment you will analyze different foods and determine their calorie content. To do this, you will use a calorimeter. In a calorimeter, food is burned and the amount of heat released by the burning is measured. In the experiment, you

> Fats provided 9 Calories per gram, whereas proteins and carbohydrates provide only 4 Calories per gram.

> The **calorie** is a measure of energy. In chemistry a calorie (cal) is equal to the energy required to increase the temperature of 1 gram of water by 1°C. A food Calorie (Cal) is equal to 1000 cal or 1kcal.

will use the heat released from the food to heat a known mass of water, and then measure the increase in temperature.

Heat content is dependent not only on the quantity of a substance but also on the identity of the substance. Each substance has its own specific heat. The specific heat of water is 1.0 calorie/g°C. This statement means it takes 1.0 calorie of energy to raise the temperature of 1.0 g of water by 1 degree Celsius. Ethyl alcohol has a specific heat of only 0.59 calories/g°C. Much less energy is needed to raise the temperature of 1 g of ethyl alcohol by 1°C.

To calculate the heat given off when you burn the food, there is a third aspect to consider (besides the mass and the specific heat of the material): temperature change (ΔT). The calculation is expressed as

Heat given off or absorbed = (specific heat)(mass)(temperature change)

So you will need to monitor the temperature change and know the mass of the water to do this calculation.

MATERIALS

Equipment
calorimeter

thermometer

graduated cylinder

matches

Chemicals
various foods

> A **calorimeter** is an apparatus used to measure heat.

PROCEDURE

1. Set up the calorimeter. The calorimeter consists of a flask that can be suspended above the burning food and a can with a opening at the bottom to allow the food sample to be placed uunder the flask as well as allow oxygen into the can.

2. Weigh the flask from the calorimeter. Add 50 mL of water to the calorimeter flask and reweigh. Record this mass in the Data section.

3. Allow the water to stand for several minutes to reach a stable temperature. Record the temperature; this is the initial temperature (T_i). Leave the thermometer in the flask to monitor the temperature throughout the experiment.

4. Obtain a sample of food that is approximately 0.3–0.5 grams. Weigh the food and record the exact mass and food type in the Data section.

5. Secure the food sample on the pin that is connected to a cork support. Ignite the food and slide the burning food under the calorimeter flask.

6. When the food has finished burning, recored the the final temperature (T_f) in the Data section.

7. After the food has cooled, weigh any food left on the pin plus any ashes you collect. This will be subtracted from the initial weight to get the actual mass that was burned.

8. Once you have collected all the data for one food sample, obtain a second food sample and run the experiment again. Be sure to replace the water because

the previous water has been heated and will not remain at a steady initial temperature. As you did earlier, allow the water to come to room temperature and record that temperature in the Data section.

9. Complete Steps 1–6 for the new food sample.

Name: _____

DATA AND RESULTS

Name of food product analyzed:

Sample 1: _____

Sample 2: _____

1. Mass of food used:

	Sample 1	Sample 2
Mass of flask + water	_____	_____
Mass of flask (-)	_____	_____
Mass of water (=)	_____	_____

2. Mass of food burned:

	Sample 1	Sample 2
Initial mass of food	_____	_____
Mass of food after burning (-)	_____	_____
Mass of food burned (=)	_____	_____

3. Change in temperature of water:

	Sample 1	Sample 2
Final temperature, T_f	_____	_____
Initial temperature, T_i (-)	_____	_____
Temperature change, ΔT (=)	_____	_____

4. Calculate the calories released by each sample.

   Heat released = (specific heat of water)(mass food)(temperature change, ΔT)

   Sample 1:

   Sample 2:

QUESTIONS

1. Which food sample released the most energy?

2. Since a calorie is a measure of the heat released by burning food, do you believe that your results are an accurate representation of the amount of calories in your food sample? Explain.

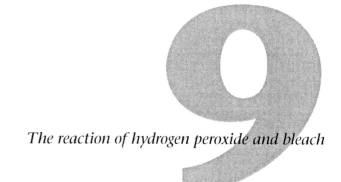

The reaction of hydrogen peroxide and bleach

STOICHIOMETRY and the CHEMICAL REACTION

You would be surprised at the number of chemical reactions you can do with common household products. In fact, you need to be very careful when working with some cleaning agents because they can react to form toxic gases. In this experiment, you will be mixing two common household products, but the product in this case is a nontoxic gas, oxygen. When hydrogen peroxide (H_2O_2) is mixed with bleach, bubbles of oxygen gas are formed. The active ingredient in bleach is sodium hypochlorite, NaOCl, which is present as a 5.25% solution in water. In this experiment, you will measure how much oxygen is produced when known amounts of hydrogen peroxide solution are mixed with known amounts of bleach. Based on the amount of reactants used (bleach and hydrogen peroxide) and the amount of product formed (oxygen), you will determine the stoichiometry of the chemical reaction.

If you mix bleach and ammonia together, the toxic gas chlorine can form.

Stoichiometry is the study of the quantitative relationship between substances undergoing chemical reactions. You may have encountered similar sorts of problems outside the lab—for instance, figuring out how many cookies (products) can be made from a given amount of ingredients (reactants) based on a specific recipe (equation). If a recipe requires that 2 cups of flour be mixed with four eggs to make 24 cookies, how many cookies can be made from 1 cup of flour and two eggs? Like the ratio of the ingredients in cooking, the ratio of the reactants in a chemical reaction are fixed. So how many cookies can you make if you have 1 cup of flour and a dozen eggs? It is obvious that the number of eggs is in excess, or you might say that the number of

The **limiting reagent** restricts the amount of product formed.

cookies you can make is limited by the amount of flour. In chemistry, the reagent that is completely used up in a reaction is called the **limiting reagent.**

The reaction between bleach and hydrogen peroxide acts as if it were following a recipe. How will the volume of oxygen produced from 4 mL of hydrogen peroxide vary with the amount of bleach added to it? Make a prediction. Did your reasoning include the fact that the amount of oxygen gas produced is dependent on both the amounts of hydrogen peroxide and bleach? These two compounds react according to specific mole ratios that you will be determining from the results of this experiment. You will react a fixed amount of bleach with varying amounts of peroxide. At some point, no matter how much peroxide you add, you will always get the same amount of oxygen. This point is when the bleach is the limiting reagent—4 mL of bleach can only make a certain amount of oxygen, just like you can have a dozen eggs but, if you only have 1 cup of flour, you will only get so many cookies. You will repeat this process by maintaining a constant amount of peroxide and vary the bleach. This will give you the limiting amount of peroxide. With this information, you can write a balanced equation for the reaction.

MATERIALS

Equipment
large beaker for waste collection

small beakers (2)

100-mL graduated cylinder

10-mL graduated cylinder (2)

tweezers

small vial

rubber tubing and stopper

250-mL side-arm Erlenmeyer flask

ring stand with clamp

water trough

Chemicals
3% aqueous hydrogen peroxide, H_2O_2

bleach, 5.25% NaOCl

PROCEDURE

You will run two sets of reactions for this experiment as stated above. In Reaction Set A, the volume of bleach will remain constant (4 mL) and the volume of hydrogen peroxide will vary (table labeled Set A).

Set A	Volume of bleach	Volume of H_2O_2
Run 1	4 mL	1.5 mL
Run 2	4 mL	2.5 mL
Run 3	4 mL	3.5 mL
Run 4	4 mL	4.5 mL
Run 5	4 mL	5.5 mL
Run 6	4 mL	6.5 mL
Run 7	4 mL	7.5 mL

In Reaction Set B, the volume of hydrogen peroxide will remain constant (4 mL) and the volume of bleach will vary. The volumes of reactants to be used for each set of reactions are given below. This may seem like a lot of runs but each run takes just minutes to perform.

Set B	Volume of bleach	Volume of H_2O_2
Run 1	1.5 mL	4 mL
Run 2	2.5 mL	4 mL
Run 3	3.5 mL	4 mL
Run 4	4.5 mL	4 mL
Run 5	5.5 mL	4 mL
Run 6	4.5 mL	4 mL
Run 7	7.5 mL	4 mL

1. Fill a water trough with tap water. Completely immerse a 100-mL graduated cylinder (if necessary remove the plastic bottom from the cylinder) in the water trough, filling it with water. Turn the cylinder upside down, keeping the mouth below the surface of the water in the trough. Clamp the cylinder onto a ring stand, positioning it high enough to allow you to slip a piece of rubber tubing into the mouth of the cylinder. The up-ended graduated cylinder should be full of water (NO AIR BUBBLES) and securely fastened onto the ring stand. This is the oxygen-measuring vessel; gas formed by the reaction will bubble into the up-ended cylinder, where it will displace some of the water. Then, you will be able to read the volume displaced directly from the graduations on the cylinder. The overall setup for the gas-collecting apparatus is shown below.

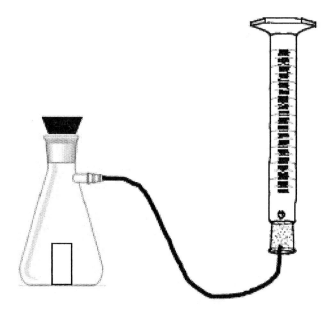

2. Label two clean, dry small beakers, one for bleach and one for hydrogen peroxide. Obtain approximately 45 mL of bleach and 45 mL of hydrogen peroxide.

3. Label the two 10-mL graduated cylinders, one for bleach and one for hydrogen peroxide. With the aid of a disposable pipette, transfer the designated amount of bleach (4 mL for Set A, reaction 1) into the 10-mL graduated cylinder. Because this experiment is to be quantitative, it is necessary to measure the quantities for each run exactly. Pour the bleach into the side arm Erlenmeyer flask.

4. Measure out the designated amount of hydrogen peroxide (1.5 mL for Set A, reaction 1) into a 10-mL graduated cylinder. Pour the hydrogen peroxide into the small vial, then use tweezers to lower the vial into the sidearm flask, taking care not to knock over the vial.

5. Again being careful not to knock over the vial, stopper the flask with the rubber stopper. Push it in firmly to form a good seal. Place the loose end of the tubing into the mouth of the up-ended graduated cylinder (refer to the diagram on the previous page). You may have to hold the tubing to prevent it from flopping around. Don't worry if a bubble or two escapes from the tubing into the cylinder.

6. Once the setup is complete, jostle the reaction flask until the vial tips over and spills the hydrogen peroxide into the bleach. Swirl the flask to ensure complete mixing. Oxygen should begin forming immediately. Some reactions may finish quite rapidly; others may take several minutes. Wait until the mixture in the flask stops fizzing and oxygen stops bubbling into the graduated cylinder. Record the amount of gas that was produced by noting the level of the liquid gas interface in the graduated cylinder. Remember that the cylinder is upside down when you read the markings. The volume of oxygen produced should be recorded in the data table in the Data section.

7. Empty the flask by pouring the reaction solution into the large beaker that will serve as your waste container. Rinse the flask and vial several times with water. The flask and the vial do not have to be dry for the subsequent runs.

8. Refill the graduated cylinder with water for the next run. Repeat all of the reactions noted above for Reaction Set A and Reaction Set B. Record the volume of oxygen produced in each run.

A look at the controversial compound caffeine

EXTRACTION of CAFFEINE from BEVERAGES

Few compounds consumed by Americans are surrounded by as much controversy as caffeine. One article tells us that caffeine will cause anxiety, whereas another article tells us that the consumption of a cup of coffee before an exam will improve our grade. Approximately 50% of Americans drink coffee daily at a rate of over three cups per day. Caffeine is also found in soda pop, tea, chocolate, and pills that keep you awake. In addition to caffeine, tea and chocolate contain a related compound, theobromine. While the effects of theobromine are mild when compared to caffeine, it is believed the effects of theobromine are enhanced by the presence of the caffeine in chocolate. The structures of caffeine and theobromine are shown below.

Amines that are found in plants are called **alkaloids**.

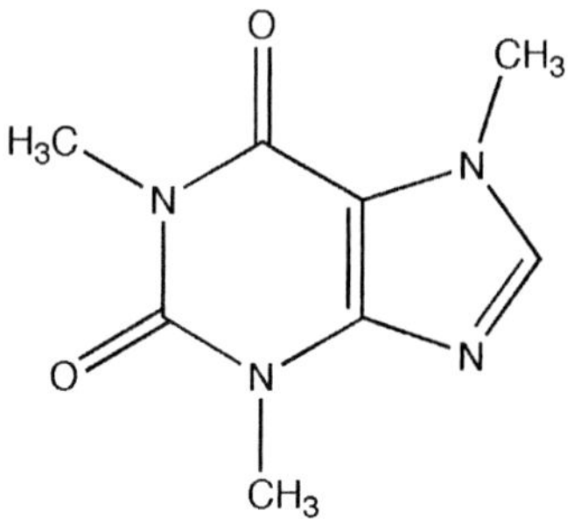

Caffeine

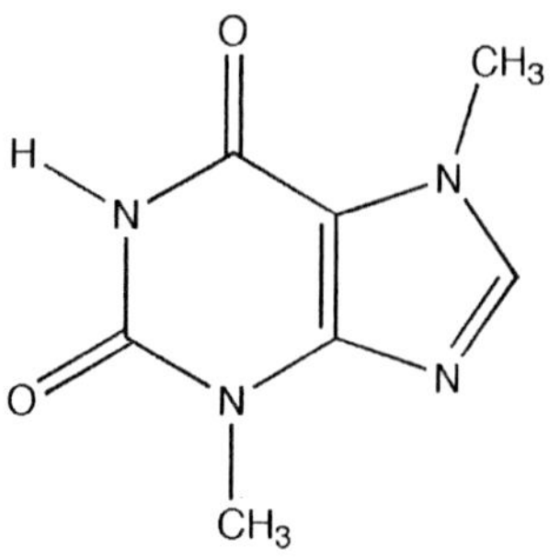

Theobromine

Caffeine is one of the oldest known stimulants. Caffeine stimulates both the central nervous system and skeletal muscles. This stimulation results in increased alertness, which is believed to be the reason for increased ability to concentrate, in the short term. The increased alertness also aids in the ability to stay awake when tired. The down side to this effect is that excessive intake of caffeine may result in restlessness, insomnia, and irritability. Heavy caffeine use can lead to both a tolerance to and a dependence on the consumption of caffeine. Anyone who has tried to kick the caffeine habit can tell you about the headache from caffeine withdrawal. The symptoms are real and, in extreme cases, can be so severe that the symptoms can include nausea and lethargy. Many headache remedies include caffeine as one of the components. It is believed that the addition of caffeine makes the pain reliever more effective.

Check your caffeine consumption by examining the list of common caffeine-containing products below. How much caffeine do you consume in a day?

Product	Caffeine content/serving
Coffee (7 oz)	90–150 mg
Espresso (2 oz)	100 mg
Stay-awake pill	100 mg
Instant coffee	60–80 mg
Tea	30–70 mg
Cola	30–45 mg
Chocolate bar	30 mg
Cold-relief tablet	30 mg
Chocolate milk (8 oz)	8 mg

In this experiment, you will extract the caffeine from some typical caffeine-containing products. The extraction will be a crude one, but it will give you an indication of the varying amounts of caffeine found in different products. You may also be asked to look at the purity of your isolate.

You can analyze the purity of your isolate caffeine by thin-layer chromatography (TLC). TLC is often used in organic chemistry to identify the components of a mixture. In TLC, a small drop of the mixture to be separated is applied near one end of a flexible plate coated with silica gel. The end of the plate is then placed into a developing solvent, called the **mobile phase,** which flows up the plate by capillary action. As the developing solvent flows up the plate, it can carry along the components of the mixture. The rate at which a particular component moves depends on whether it tends to dissolve in the developing solvent or remain adsorbed (or stuck) on the surface of the silica gel, called the **stationary phase.** A component that moves rapidly is spending more time dissolved in the developing solvent, whereas a component that moves slowly is spending more time adsorbed to the silica. The TLC plate is then analyzed under UV light to see if any organic compounds are present.

The caffeine in your isolate sample should move with the same speed as a reference sample of pure caffeine. If any additional components (impurities) are present in your isolate, you will see other spots present on the TLC plate.

MATERIALS

Equipment
125-mL separatory funnel
250-mL beaker (2)
125-mL Erlenmeyer flask
50-mL Erlenmeyer flask
mortar and pestle
100-mL graduated cylinder
spatula
funnel
test tubes
watch glass
TLC plates
UV lamp
microcapillary tubes

Chemicals
tea bags
A variety of caffeinated sodas
instant coffee
No-Doz™ tablets
methylene chloride, CH_2Cl_2
saturated aq. NaCl
calcium carbonate, $CaCO_3$
anhydrous magnesium sulfate, $MgSO_4$
ethyl acetate

PROCEDURE
Part A: Isolate of Caffeine

1. You will need 80 mL of beverage for the experiment. Choose one of the possible caffeine sources available in the lab. For the coffee and tea, make up the equivalent of a strong cup of tea or coffee and filter or decant to separate liquid from solids. For the soda pop, measure out 80 mL. Crush the No-Doz™ tablet, using the mortar and pestle, and dissolve the powder in 80 mL of water.

2. Measure 80 mL of cooled sample into a beaker, add 2 g of calcium carbonate to your sample, and swirl to dissolve.

3. Pour your sample into a 125-mL separatory funnel. Make sure the stopcock is closed.

4. Add 20 mL of methylene chloride (CH_2Cl_2) to the separatory funnel, place the stopper in the opening, and gently rock the funnel back and forth. Do not shake vigorously or an emulsion may form.

5. Place the separatory funnel in a support ring and allow the layers to separate. Remove the stopper and drain off the lower organic layer (methylene chloride plus caffeine) into an Erlenmeyer flask. If any emulsion is present, allow that to flow into the Erlenmeyer flask.

6. Repeat steps 4 and 5 and combine the second organic layer with the first in the Erlenmeyer flask.

7. If you have a significant amount of emulsion, perform step 8. If you do not have an emulsion, proceed to step 9.

8. To remove an emulsion, you will need to wash the organic layers with a salt-water solution. To do this, discard the water in the separatory funnel and rinse the funnel with water before returning the organic layer to the separatory

Separatory funnel

An **emulsion** forms when insoluble droplets of one liquid form in another liquid. If you have an emulsion you will not see a clear separation between the organic and aqueous layer.

funnel. After the funnel has been rinsed, return the combined organic layers into the separatory funnel and add 20 mL of salt water. Rock the funnel gently to mix the layers and then allow the layer to separate. Drain the lower organic layer into a clean Erlenmeyer flask.

9. Add a small scoop of anhydrous $MgSO_4$ to the organic layer. The magnesium sulfate is used to absorb any residual water in the solution. If all the $MgSO_4$ clumps, you may need to add more. You want to have some free-flowing magnesium sulfate present. Stopper the flask and allow the solution to stand for 10 minutes.

10. Set up a filter funnel with folded filter paper. Filter the solution into a clean tared 50-mL Erlenmeyer flask. Rinse the flask with a small amount of methylene chloride and pour this through the filter.

11. Evaporate the methylene chloride in the hood. The evaporation can be accomplished by using a water bath on a hotplate set to a low heat setting. When the solution gets to about 1/16 inch, remove from the water bath and let the rest evaporate in the hood. Another method for evaporation is to blow a stream of air carefully into the solution.

12. Reweigh the Erlenmeyer flask and determine the amount of caffeine.

Part B: Analysis of the Purity of the Isolated Caffeine.

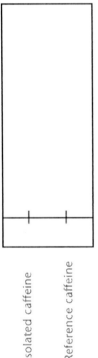

1. Obtain a TLC plate, handle the plate carefully by the edges, and do not bend. Using a lead pencil (not a pen), *lightly* draw a line across the plate (about 1 cm from the bottom). Using a centimeter ruler, lightly mark off two intervals equally spaced on the line. These are the points at which the samples, your isolated caffeine and reference caffeine, will be spotted.

2. Transfer several crystals of your isolated caffeine to a small, labeled test tube.

3. Add 5 mL of methylene chloride (CH_2Cl_2) to the test tube and swirl gently to dissolve the crystals.

4. Spot the solution of your isolate on the plate. To spot the plate, dip a capillary tube into the solution. Lightly touch the capillary tube to the plate. Allow just a small spot of liquid onto the plate.

5. Next, spot the caffeine reference solution on the plate.

6. When the plate has been spotted, it is ready to be placed in a development chamber. For a development chamber, you will use a 250-mL beaker lined with a piece of filter paper and covered with a watch glass or aluminum foil. The purpose of the filter paper is to provide an atmosphere in the developing chamber that is saturated with solvent.

7. When the development chamber has been prepared, obtain a small amount of the development solvent (ethyl acetate). Fill the chamber with the development solvent to a depth of about 1/4 inch. Be sure that the liner is saturated with the solvent. The solvent level must not be above the spots on the plate or the samples will dissolve off the plate into the reservoir instead of developing. Place the spotted plate in the chamber and allow the plate to develop.

8. When the solvent has risen to a level about 1 cm from the top of the plate, remove the plate from the chamber and, using a lead pencil, mark the position of the solvent front. Let the plate dry.
9. When the plate is dry, observe it under a short-wavelength UV lamp. Lightly outline all the observed spots with a pencil. Also, observe under ordinary light and circle any spots.

Name: _____

DATA AND RESULTS

Part A: Isolation of Caffeine

Sample chosen (beverage, brand): _____

Amount of sample used: _____

1. Determine the amount of caffeine isolate.

   Weight of flask with caffeine: _____

   Weight of empty flask: _____

   Weight of isolated caffeine: _____

2. Describe the appearance of the isolated caffeine.

Part B: Analysis of the Purity of the Isolated Caffeine

1. Draw a diagram of the chromatogram.

2. Record any observations about the appearance of the spots under ordinary light and UV illumination.

QUESTIONS

1. How did your results compare with the expected caffeine amounts listed in the table? Can you account for any difference?

2. From the appearance of your isolated caffeine, do you believe that your sample was pure caffeine? If not, what do you think are the possible contaminants?

3. Did the TLC provide you with any additional information about the purity of your isolated caffeine? Explain.

Many flowers, fruits, and vegetables contain organic compounds that change color with pH.

USING VEGETABLE INDICATORS to DETERMINE pH

Acids and bases were originally classified by their physical properties such as taste: acids tasting sour and bases tasting bitter. It was also noticed that many natural substances contained pigments that changed color when exposed to acids and bases. One of the earliest methods for determining the pH of a solution was to use chemical compounds that are derived from plants and change color with the pH of the solution. These substances are called **indicators.** One such indicator, litmus, will turn blue in base and red in acid.

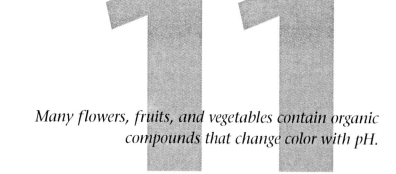

Hydrangeas will be blue or pink, depending on the acidity of the soil.

In the 1920s a more systematic definition of acids and bases was proposed by Brønsted and Lowry. They defined an acid as a substance that is capable of donating a proton; conversely, a base is a substance that accepts protons. An example of this is shown below:

$$HCl + H_2O \rightarrow H_3O^+ + Cl^-$$

In this example, hydrochloric acid, HCl, is the acid and donates a proton (H$^+$) to the base, H$_2$O, water. Hydrochloric acid is defined as a *strong* acid; in other words, it is an excellent proton donor. When hydrochloric acid is mixed with water, the above reaction essentially goes to completion, with all the HCl being converted to H$_3$O$^+$, hydronium ion, and Cl$^-$, chloride ion.

Acetic acid (which is found in vinegar) is said to be a *weak* acid:

$$CH_3COOH + H_2O \leftrightarrows CH_3COO^- + H_3O^+$$

pH = –log[H_3O^+]—the brackets indicate that the amount of hydronium ion is stated in molarity (**M**).

A weak acid is not a good proton donor, so when acetic acid is mixed with water, few hydronium ions are formed. If equal molar amounts of acetic acid and hydrochloric acid are put into separate containers of water, the solution of acetic acid has a low concentration of hydronium ions, whereas the solution of hydrochloric acid has a high concentration of hydronium ions.

To measure the strength of an acidic solution, we measure the concentration of H_3O^+. The scale used to denote the acidity of a solution is **pH.** The pH of a solution is equal to –log[H_3O^+]. Because this log is negative, a high concentration of H_3O^+ corresponds to a low pH. Thus, a 0.1 molar (0.1 M) solution of HCl, a strong acid, has a pH of 1 (–log 1×10^{-1}). A 0.1 M solution of acetic acid, a weak acid, has a pH of 3, (–log 1×10^{-3}). This pH is higher because, of the 0.1 moles of acetic acid present in each liter of solution, only 0.001 moles of the total amount has reacted to form hydronium ion. This means in there is a significant quation of unreacted (undissociated) acetic acid in the solution.

In pure water, there is a slight dissociation of the water molecules into hydronium ions and hydroxide ions:

$$2\,H_2O \rightleftharpoons H_3O^+ + OH^-$$

In pure water, the concentration of hydronium ions is 1×10^{-7} M, and because every time a H_3O^+ is formed an OH^- is formed, the concentration of OH^- is also 1×10^{-7} M. Pure water has a pH of 7 (–log 1×10^{-7}). Because the concentration of OH^- = H_3O^+, pure water is neither acidic nor basic but is considered neutral.

Most familiar bases are ionic compounds that contain a hydroxide ion, such as sodium hydroxide (NaOH) or potassium hydroxide (KOH). These bases accept protons from acids to form water:

$$H_3O^+ + OH^- \rightleftharpoons 2\,H_2O$$

Again, a strong base such as NaOH will completely dissociate in water to give a high concentration of hydroxide ions. Some bases do not contain hydroxide ions themselves but form hydroxide ions when in aqueous solution through a reaction with water. Ammonia (NH_3) is an example of such a base:

$$NH_3 + H_2O \rightleftharpoons NH_4^+ + OH^-$$

The pH system is also used to measure the strength of a basic solution. How can this be done when pH is determined by the concentration of hydronium ions but a base usually involves hydroxide ions? Remember that the concentrations of hydronium ion and hydroxide ion in pure water are both equal to 1×10^{-7} M. The product of the concentration of hydronium ion and hydroxide ion is called the ion product of water, K_w.

$$K_w = [H_3O^+][OH^-] = [1 \times 10^{-7}][1 \times 10^{-7}] = 1 \times 10^{-14}$$

The ion product of water is a constant (at constant temperature) and is the same for all aqueous solutions. Because this parameter is constant, when the concentration of hydronium ion increases, the concentration of hydroxide ion must decrease and *vice versa*. When a base is present, the concentration of hydroxide ion increases and the concentration of hydronium ion decreases. Because the pH is –log[H_3O^+], when the concentration of hydronium ion decreases, the pH increases.

Just as there are strong and weak acids, there are also strong and weak bases. In a 0.1 M solution of NaOH, the *strong* base is completely dissociated; therefore, the [OH^-] = 0.1 = 10^{-1}. Since the product of [OH^-] and [H_3O^+] must be 10^{-14} for all aqueous solutions, [H_3O^+] for this basic solution must equal $10^{-14}/10^{-1}$, or 10^{-13}. The pH of a 0.1 M solution of NaOH is thus $-\log[10^{-13}] = 13$. In a 0.1 M solution of ammonia (NH_3), a weak base, only about 1/100 of the ammonium hydroxide is dissociated at any point in time so [OH^-] = 0.1/100 = 10^{-3}. The pH of a 0.1 M ammonia solution is $-\log[10^{-14}/10^{-3}] = -\log[10^{-11}] = 11$.

How does one determine the pH of a solution? One can use plant indicators, like those previous mentioned. Some indicators have a wide range of color changes. In this experiment, you will extract colored substances from red cabbage and use them as an indicator. First, you will determine the color the indicator will be at a specific pH, using buffer solutions of known pH provided for you in the lab. Then, you will take various household substances and common laboratory solutions and determine the pH of their aqueous solutions.

MATERIALS

Equipment
100-mL graduated cylinder
500-mL beaker
test tubes and rack
eyedropper
grease pencil
hotplate
mortar and pestle

Chemicals
red cabbage
buffer solutions
various household chemicals
0.1 M HCl
0.1 M acetic acid
0.1 M ammonium hydroxide

CAUTION:
In case of spills, wash your skin thoroughly with water and clean up the lab bench.
Use caution with the household chemicals. Familiarity tends to breed carelessness.

PROCEDURE

1. Place several purple cabbage leaves in a 500-mL beaker and cover the leaves with water. Boil the cabbage leaves to remove the pigment.

2. While the indicator cools, set up an array of buffer solutions of different pHs. Label ten small test tubes to correspond with the pHs of the buffer solutions (pHs 2 to 11). Fill each tube approximately one-half full (~5 mL) with the appropriate buffer from the stock solution bottle.

3. Arrange the tubes in order of increasing pH values in your test-tube rack. Add several drops of the cabbage indicator to each tube with an eyedropper. You should get a nice array of colors from the vegetable dye. Record the color of each pH. It's fine to add more of the cabbage indicator to intensify the color but be sure to add the same amount to each test tube.

4. Test the pH of the household products available in the lab by placing a 5-mL sample in a test tube and adding cabbage indicator. Solid samples such as vitamin C or antacid tablets should be prepared by crushing about one-quarter of the tablet and dissolving it in ~5 mL of deionized water. Compare the resulting colors with your buffer array to determine an approximate pH.

5. With the cabbage indicator, test the pH of several laboratory reagents.

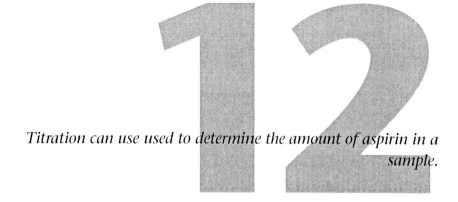

Titration can use used to determine the amount of aspirin in a sample.

TITRATION of ASPIRIN

Analytical chemistry is the science of making quantitative measurements. Besides performing reactions, chemists are often called on to determine the identity and/or amount of a chemical in a sample. For example, forensic chemists can be called upon by drug enforcement personnel to determine the identity of an unknown "white powder." Or, the FDA might want to ensure that the correct dosage of a drug is found within an over-the-counter drug. There are many analytical methods used by chemists; one of them is titration. **Titration** is the quantitative measurement of a compound in solution done by reacting it completely with a reagent. One type of titration is an acid-base titration.

Quantitative measurements determine the quantity of each ingredient of a substance. Qualitative analysis determines the constituents of a substance without regard to the quantity of each ingredient.

An acid-base titration utilizes the neutralization reaction between an acid and a base. An example of an acid-base reaction is the reaction of an aqueous solution of hydrochloric acid, HCl, with an aqueous solution of sodium hydroxide, NaOH.

$$HCl\ (aq) + NaOH\ (aq) \rightarrow NaCl\ (aq) + H_2O\ (l)$$

A **neutralization reaction** is a reaction in which a proton is transferred from an acid to a base, producing a salt and, in some cases, water.

The products of this neutralization reaction are a salt and water. Because NaOH consists of Na^+ and OH^- ions and the HCl ionizes in water to give H_3O^+ and Cl^- ions, we may rewrite the above equation as

$$H_3O^+ + Cl^- + Na^+ + OH^- \rightarrow Na^+ + Cl^- + 2\ H_2O$$

If we ignore the ions that appear on both sides of the equation, the resulting equation is

$$H_3O^+ + OH^- \rightarrow 2\ H_2O$$

Ions that appear on both sides of the equation are called **spectator ions.**

The balanced chemical equation tells us the number of moles of base required to completely react with a given number of moles of acid. Titration is a common

A **buret** is an apparatus for delivering measured quantities of liquid. It consists essentially of a graduated glass tube, usually furnished with a small aperture and stopcock (see below).

1 gram = 1000 mg

laboratory procedure used to determine the concentration of a solution of base (or acid) by reacting the base with a known amount of acid (or base) of known concentration.

To perform a titration, a known amount of an acid is placed in a flask. Then base is added from a buret until just enough has been added to completely react with the acid. This point can be determined by use of an indicator that changes color when the reaction is complete. The point at which the indicator changes color is called the **end point** of the titration.

In this experiment, you will determine the amount of aspirin (acetylsalicylic acid) in an aspirin tablet or in a sample made in the laboratory. You will place a sample of aspirin into a flask and titrate the sample with a base, NaOH, of known concentration. You will be verifying the amount of aspirin in a tablet by comparing it to the amount given on the label or you will be checking the purity of a sample synthesized in the lab. The reaction for aspirin with sodium hydroxide is given below.

$$C_9H_8O_4(aq) + NaOH(aq) \rightarrow C_9H_7O_4^- + Na^+ + H_2O$$

Aspirin   Sodium hydroxide

How do we determine the amount of aspirin by titration? The amount of the aspirin in this experiment can be calculated from the titration data by using the fact that at the end point, the moles of aspirin and NaOH are equal. During the titration, you add NaOH until the indicator changes color. The volume of base added contains a number of moles equal to the number of moles of aspirin. If we know the concentration of the base, we can calculate the number of moles of aspirin in the sample by using the following equation.

(molarity NaOH) × (volume NaOH) = moles of aspirin

Remember, molarity (M) is equal to moles/liter. If it takes 20.0 mL (0.0200 L) of 0.125M NaOH to reach the endpoint of the titration then

0.0200 L × 0.125 moles/L = 0.00250 moles NaOH

There was 0.00250 moles of NaOH in the volume you titrated, therefore there was 0.00250 moles of aspirin in the flask. Now you need to convert the number of moles of aspirin to a mass. The molecular weight of aspirin is 180.16 grams/mole. If you have 0.0125 moles of aspirin, then

0.00250 moles × 180.16 grams aspirin /mole = 0.450 g aspirin

If you were analyzing a tablet that was said to contain 500 mg of aspirin, do you think you got your money's worth?

This lab is *quantitative*, so careful attention to technique is necessary. The indicator you will use is phenolphthalein. It is colorless in acidic solutions and pink in basic solutions. At the *exact* end point the solution is neutral; you have added an amount of basic solution that contains the number of moles equivalent to the moles of acid. At one drop past the end point the base is in excess and this excess will make the solution basic (pink). You want to have a light pink color at the endpoint, dark pink indicates that you have added too much base.

MATERIALS

Equipment
10-mL graduated cylinder

small beakers

125-mL Erlenmeyer flask

mortar and pestle

ring stand and buret clamp

stir bar

stir plate

50-mL buret

Chemicals
standardized aqueous NaOH (0.1-0.2 M)

aspirin samples

phenolphthalein indicator solution

ethanol

PROCEDURE

1. Use a buret clamp to secure the 50-mL buret to a ring stand. Thoroughly rinse the 50-mL buret with deionized water. The deionized water should drain evenly from the inside surfaces of the buret and leave no droplets of water behind. Any droplet formation on the inside of the buret means that the buret is dirty and needs to be cleaned.

   CAUTION: A buret is used to perform the titration. Burets are fragile and expensive, so take care!

2. Obtain about 50 mL of sodium hydroxide solution (~0.1 M to 0.2 M) in a small beaker. This solution has been standardized, so be sure to write down the exact concentration of this solution

3. Fill the buret with sodium hydroxide solution. Place a small beaker under the buret and open the stopcock long enough to fill the tip with solution and remove any air bubbles. Add solution to the top of the buret, or drain the solution through the stopcock, to bring the level close to 0.00 mL on the buret scale. It is not necessary that the initial level is exactly 0.00 mL, but it must be at or below the start of the scale.

   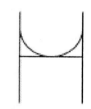
   The **meniscus** is the curved upper surface of a column of liquid.

4. Record the initial level of the buret, keeping your eye level with the bottom of the meniscus. Estimate the last digit.

5. Prepare your aspirin sample.

   a) If you are analyzing an aspirin tablet you will need to crush and grind the tablet in a mortar. Once the tablet is a fine powder, transfer the powder to an Erlenmeyer flask. Rinse the mortar with a small amount of ethanol to remove all the sample. Be sure to write the amount of aspirin the manufacturer states is in the aspirin tablet.

   b) If you are testing a sample made in the lab, weigh out a 0.200 g sample . It does not have to be exactly 0.200 grams, but you do need to record the exact amount of the sample in the Data section.

6. Add approximately 10 mL of ethanol to the flask. Aspirin is only slightly soluble in water so ethanol will help to get the aspirin into solution.

7. Add approximately 15 mL of water to the flask and add the stir bar. Place the flask on the stir plate and stir the sample for about 5 minutes. If you have used an aspirin tablet, there are filler present which may not dissolve.

8. Add several drops of phenolpthalein indicator to the flask.

9. Titrate the aspirin solution to the end point, the first permanent appearance of pink in the solution. Add the base rapidly at first as you continuously stir the flask and its contents to ensure rapid mixing. As you approach the end point you will see flashes of pink in the solution, which disappear upon stirring. At this point continue the addition of base more slowly — eventually drop by drop — mixing carefully after each addition. With practice, it is possible to determine the end point with the precision of one drop.

10. Record the final volume in the buret.

11. Dispose of the solution from your titration in the waste container. Rinse your flask with water and repeat the titration with another aspirin sample. Check the volume of sodium hydroxide that you have left in the buret. You may wish to refill the buret with more solution before beginning the second titration.

12. If the volumes of NaOH delivered for the two trials are not within reasonable agreement, do a third titration.

> *I believe that you can, by taking some simple and inexpensive measures, extend your life and your years of well-being. My most important recommendation is that you take vitamins every day in optimun amounts, to supplement the vitamins you receive in your food.*
>
> *Linus Pauling, Ph.D., two-time Nobel Prize laureate*

VITAMIN C CONTENT in TABLETS and BEVERAGES

Linus Pauling won a Nobel Prize for chemistry in 1954 and then for peace in 1962. He was also a great proponent of the use of vitamin C to prevent anything from colds to cancer. While there is little doubt of his significance in chemistry, there are many who do not agree with his views on the benefits of vitamin C. Yet, how many people do you know who insist that a dose of vitamin C will cure or prevent a cold? Whether you agree with Pauling's statement or not, vitamin C is a necessary requirement for a healthy diet.

Pure vitamin C (also called ascorbic acid) is a white, crystalline organic compound with the chemical formula $C_6H_8O_6$. The structure of ascorbic acid is shown below:

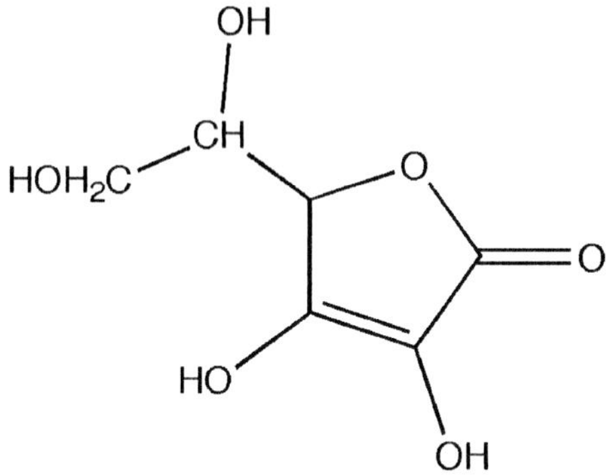

Vitamin: an organic substance essential in small quantities to the nutrition and normal metabolism of most animals.

Vitamin C is a weak acid, is very soluble in water, and thus is considered one of the water-soluble vitamins. One of the most important chemical properties of ascorbic acid is that it is a reducing agent, that is, it is relatively easily oxidized.

Ascorbic acid is widely distributed in both the plant and animal kingdom. It is synthesized by all known species except the primates (including man), the guinea pig, Indian fruit bats, and the red-vented bulbul bird — rather a select group. These species are unable to synthesize ascorbic acid because their cells do not have the enzyme that catalyzes the reaction to form L-ascorbic acid. A vitamin C dietary deficiency can result in scurvy, which in primates first appears as symptoms of weakness and fatigue followed by shortness of breath and aching in the bones.

Although we generally associate vitamin C with citrus fruits like oranges or orange juice, vitamin C can be obtained from a variety of sources. American Indians knew that pine needle broth and rose hip tea, which are both good sources of vitamin C, would cure scurvy. Eskimos traditionally eat certain layers of whale skins to obtain vitamin C. Today the most common dietary source of vitamin C is the vegetable-fruit group of foods. It is important to remember that the two chemical characteristics discussed earlier, water solubility and ease of oxidation, can directly lead to significant vitamin C losses during preparation. Blanching, boiling, and long-term storage of vegetables can all lead to substantial losses. Even slicing or maceration of fresh vegetables releases enzymes (e.g., ascorbate oxidase) that can quickly oxidize vitamin C. The keys to good vitamin C intake are fresh, quickly cooked, appropriate foods and, of course, vitamin supplements. Below is a list of the vitamin C content in some selected foods.

1 gram = 1000 mg

Food	Vitamin C content (mg)/100 g food
asparagus, cooked	27
green peppers, raw	128
onions, raw	10
parsley, raw	175
apples, raw	11
blueberries, raw	14
cranberry juice	16
oranges	37
raspberries	25
strawberries	59
rosehips	+1000

In this lab you will analyze the vitamin C content in vitamin C tablets and in other foods. The analysis will be done by titration. **Titration** is a common procedure used for determining the concentrations or amounts of substances in solutions. In a titration you add a known amount of one reactant to another reactant of unknown concentration. The reactant is added until the reaction is complete; this is called the **endpoint**. You will recognize the completed reaction by a color change or other highly visible change at the endpoint of the reaction. Knowing the reaction involved, you can calculate the unknown amount of the second reactant from the known amount of the first reactant.

In this titration, a sample of vitamin C will be titrated with iodine (I_2). See reaction below.

$$I_2(aq) + C_6H_8O_6(aq) \rightarrow C_6H_6O_6(aq) + 2\,H^+(aq) + 2\,I^-(aq)$$

Iodine + Vitamin C → Dehydroascorbic acid + Hydrogen ion + Iodide ion

In this analysis, the endpoint is signaled by the reaction of iodine with starch suspension, producing a blue-black product. As long as ascorbic acid is present, the iodine is quickly converted to iodide ion, and no blue-black iodine–starch product is observed. Once the reaction of iodine with ascorbic acid is complete, then any additional iodine will remain in solution and will react with the starch to form a blue-black color.

Because it is difficult to prepare iodine solution with an exact known concentration, you will first determine the amount of iodine needed to titrate a known concentration of ascorbic acid (mg ascorbic acid/mL iodine solution). Once you have determined this ratio you can use this conversion factor to determine the amount of ascorbic acid in a vitamin C tablet or in a beverage. Depending on the length of your laboratory period, you may do both or only one of these determinations. If you are curious about a particular food or beverage, consult with your instructor to determine if you can bring in a sample of your own.

MATERIALS

Equipment
10-mL graduated cylinder or
   10-mL volumetric pipet and bulb
125-mL Erlenmeyer flask
small beaker
ring stand and buret clamp
1-L volumetric flask
funnel
50-mL buret

Chemicals
vitamin C tablets (1000mg/tablet)
standard ascorbic acid solution (≈1mg/mL)
standard iodine solution
1% starch solution
various beverages (white grape juice, orange juice, apple juice, rose hip tea, etc.)

CAUTION:
A buret is used to perform the titration. Burets are fragile and expensive, so take care!

PROCEDURE

Part A: Preparation of Solution from Vitamin C Tablet

1. Obtain a vitamin C tablet. Weigh and record the mass to the nearest milligram (0.001 g).

2. Place the tablet in a clean 1-L volumetric flask. Fill the flask about half-full with water. Shake and swirl the flask to begin dissolving the tablet. It is not necessary at this point to shake until the tablet is fully dissolved because you will allow it to stand while you perform the next step.

The vitamin C tablet solution may already be prepared for you. Check with your laboratory instructor.

Part B: Standardization of Iodine Solution

1. Obtain approximately 30-mL of standard ascorbic acid solution. Write down the exact concentration of the solution given on the container.

2. Transfer 10 mL of the standard ascorbic acid solution into a 125-mL Erlenmeyer flask. Add 25 mL of water to the flask to give sufficient volume to allow for good mixing. Add 5–8 drops of 1% starch solution.

3. Rinse a clean buret with a small amount of iodine solution and discard into waste container. Using a funnel, fill the buret with iodine solution. Open the stopcock long enough to fill the tip with solution and remove any air bubbles. Add iodine solution to the buret to bring the level close to 0.0 mL on the buret scale. It is not necessary that the initial level is exactly 0.0 mL, but it must be at or below the start of the scale.

4. Record the initial level of the buret, keeping your eye level with the bottom of the meniscus.

5. Add the iodine slowly to the Erlenmeyer flask while you gently swirl it. The first permanent appearance of blue in the solution (after several seconds of swirling) is the endpoint. If you place a piece of white paper below the Erlenmeyer flask, this will help you to see the color change.

6. Record the final volume from the buret.

7. Repeat the titration with another 10-mL sample of ascorbic acid preparing the sample as you did in Step 2. Check the level in the buret before starting the titration to make sure there is sufficient iodine solution to complete the titration.

8. Repeat Steps 4–6.

9. If the volumes of iodine used in the two titrations are not within reasonable agreement, do a third titration.

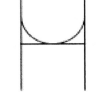

The **meniscus** is the curved upper surface of a column of liquid.

Part C: Determination of the Amount of Ascorbic Acid in a Vitamin C Tablet

Note: This solution may already be prepared for you. If so, go to Step 2.

1. While you were performing the standardization of the iodine solution, the vitamin C tablet should have dissolved in the 1-L volumetric flask. You may see on the bottom of the flask some residual powder made up of insoluble materials (most likely starch) that have been added to the tablet as filler. Add water to the flask up to the line around the neck of the flask. The bottom of the meniscus should be just touching the line. Place the cap on the flask and mix the contents thoroughly. Do this by holding the cap firmly in place and repeatedly turning the flask upside down.

2. In a clean beaker, obtain at least 30 mL of the vitamin C tablet solution.

3. Transfer 10 mL of the vitamin C tablet solution into a 125-mL Erlenmeyer flask. Prepare the sample as you did in the previous section, adding 25 mL of water and 5–8 drops of 1% starch solution.

4. If necessary, add additional iodine solution to your buret and record the initial volume. Titrate the vitamin C tablet solution as you did in Steps 4–6 of Part B. Record the final volume.

5. As before, perform at least two titrations. Perform a third titration if there is not good agreement in the volumes of the first two titrations.

Part D: Determination of the Amount of Vitamin C in a Beverage

1. In a clean beaker, obtain at least 30 mL of one of the available beverages.
2. Transfer 10 mL of the beverage into a 125-mL Erlenmeyer flask. Add 25 mL of water and 5–8 drops of 1% starch solution.
3. Repeat the titration as you have done in the previous parts, Steps 4–6. Be careful to check the volume of iodine solution in the buret and to record the initial and final volumes. In this titration you need to consider the color of the beverage when observing the endpoint. In the titration of ascorbic acid standard and vitamin C, the solution is clear but the beverage may be colored. For example, cranberry juice is a good source of vitamin C, but is highly colored. This does not mean you cannot perform the titration, it just means that the endpoint will not necessarily be blue. In the case of cranberry juice, you would more likely see a change in color from the reddish color to a darker, more purple color.
4. As before, perform at least two titrations. Perform a third titration if there is not good agreement in the volumes of the first two titrations.

Name: _____

DATA

Standardization of Iodine Solution

Concentration to ascorbic acid solution: _____ mg/mL

Volume of ascorbic acid solution: 10 mL

Ascorbic acid solution	Initial buret volume	Final buret volume	Total volume iodine solution added
Sample 1			
Sample 2			
Sample 3			

Titration of Vitamin C Solution

Volume of vitamin C tablet solution: 10 mL

Mass of vitamin C tablet: _____

Vitamin C tablet solution	Initial buret volume	Final buret volume	Total volume iodine solution added
Sample 1			
Sample 2			
Sample 3			

Titration of Beverage

Beverage used: _____

Volume of beverage: 10 mL

Beverage	Initial buret volume	Final buret volume	Total volume iodine solution added
Sample 1			
Sample 2			
Sample 3			

RESULTS

Determination of Conversion Factor (mg ascorbic acid/mL iodine solution)

If you determine the ratio of mg ascorbic acid titrated by 1 mL of iodine solution, then you can use this value to determine the amount of ascorbic acid (or vitamin C) in the other solutions in this experiment.

Volume of standard ascorbic acid solution used: 10 mL

Concentration of standard ascorbic acid solution from data section: _____ mg/mL

Average mL required to titrate a 10-mL sample of ascorbic acid: _____ mL

1. Determine mg of ascorbic acid in 10 mL ascorbic acid sample:

10 mL of ascorbic acid (AA) solution × concentration of ascorbic acid solution = mg ascorbic acid in sample

10 mL AA × $\dfrac{\text{mg AA}}{\text{mL of soln.}}$ = _____ mg AA in sample

2. Determination of conversion factor (mg AA/mL of iodine solution)

Conversion factor : _____ mg of AA in sample/_____ mL of iodine solution

3. Report conversion factor in mg of ascorbic acid/1 mL of iodine solution below:

Determination of mg Ascorbic Acid in a Vitamin C Tablet

Amount of vitamin C per tablet stated on bottle: _____ mg/tablet

Average mL required to titrate a 10-mL vitamin C tablet solution: _____ mL

1. Determine mg of ascorbic acid in 10-mL vitamin C tablet sample.

Multiply the mL of iodine solution needed to titrate your vitamin C (ascorbic acid) sample times the conversion factor from above and you will get the amount of ascorbic acid in your sample.

_____ mL of iodine soln. × $\dfrac{\text{mg AA}}{\text{mL of iodine soln.}}$ = _____ mg AA in 10-mL sample

Name: _____

2. Determine mg of ascorbic acid in vitamin C tablet.

To determine the mg of ascorbic acid in the vitamin C tablet you need to account for the fact that you only titrated a 10-mL sample out of a solution of 1000 mL.

$$\frac{\text{\_\_\_\_ mg AA}}{\text{10-mL sample}} \times \frac{1000 \text{ mL}}{1 \text{ L}} \times \frac{1 \text{ L}}{\text{tablet}} = \text{_____ mg AA/tablet}$$

The amount of ascorbic acid (vitamin C) in one tablet is _____ mg.

Determination of mg of Vitamin C in 10 mL of Beverage

Amount of vitamin C per serving stated on bottle (if given): _____ mg

Size of one serving: _____ mL

Average mL required to titrate 10 mL of beverage: _____ mL

1. Determine mg of ascorbic acid in 10 mL of beverage:

Multiply the mL of iodine solution needed to titrate your beverage sample times the conversion factor from above and you will get the amount of ascorbic acid in your beverage sample.

$$\text{\_\_\_\_\_ mL of iodine soln.} \times \frac{\text{\_\_\_\_\_ mg AA}}{\text{mL of iodine soln.}} = \text{_____ mg AA in 10-mL sample}$$

$$\frac{\text{\_\_\_\_ mg AA}}{\text{10 mL sample}} \times \frac{\text{\_\_\_\_ mL}}{\text{serving}} = \text{_____ mg AA/serving}$$

2. The amount of ascorbic acid (vitamin C) in one serving is _____ mg.

3. Report this amount and beverage type in laboratory in order to compare your results with other students.

QUESTIONS

1. How does the amount of vitamin C purported to be in the tablet compare with the amount you determined by titration? Can you given any reasons for any differences you see?

2. What fraction of the mass of the vitamin C tablet was actually ascorbic acid?

3. How does the amount of vitamin C reported to be in one serving of beverage compare with the amount you determined by titration? Which beverage was reported to have the most vitamin C? Were you surprised by the results?

4. How might the color of the beverage affect your ability to give an accurate reading in the titration? Did you have any difficulty determining the endpoint?

14

Historically, soap was made from beef tallow (fat) and wood ashes.

SYNTHESIS of SOAP

In this experiment, you will prepare soap from animal fat (lard). Animal fats and vegetable oils are esters of long-chain carboxylic (fatty) acids and the alcohol, glycerol. Glycerol contains three alcohol functional groups and thus can react with three carboxylic acids. Soaps can be formed from animal fats and vegetable oils by the hydrolysis under alkaline conditions, saponification (reaction shown below).

Fats and oils are called **triacylglycerols** or **triglycerides**.

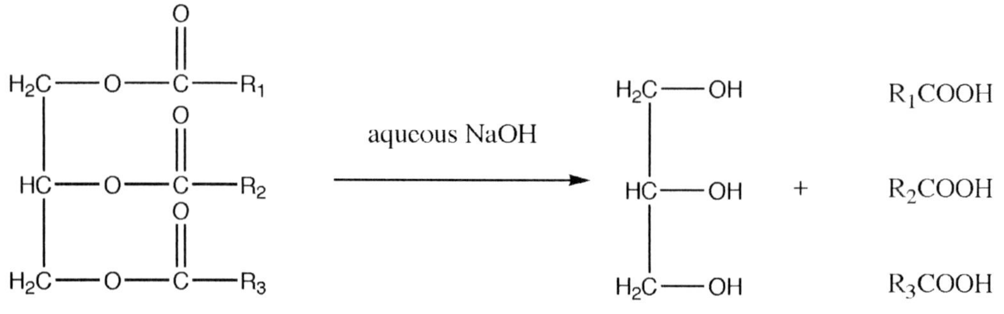

$$\begin{array}{c} H_2C-O-\overset{O}{\underset{\|}{C}}-R_1 \\ | \\ HC-O-\overset{O}{\underset{\|}{C}}-R_2 \\ | \\ H_2C-O-\overset{O}{\underset{\|}{C}}-R_3 \end{array} \xrightarrow{\text{aqueous NaOH}} \begin{array}{c} H_2C-OH \\ | \\ HC-OH \\ | \\ H_2C-OH \end{array} + \begin{array}{c} R_1COOH \\ \\ R_2COOH \\ \\ R_3COOH \end{array}$$

Triacylglycerol

The fatty acids in a triglyceride are rarely of a single type in any given fat or oil. In fact, a single triglyceride molecule may contain three different acid residues (R_1COOH, R_2COOH, R_3COOH). Each fat or oil, however, has a characteristic distribution of the various types of acids possible. For example, beef tallow generally contains 50% oleic acid, with the majority of the remaining acids being myristic (C_{14}), palmitic (C_{16}), and stearic (C_{18}) unsaturated acids.

Tallow is the principal fatty material used in making soap. Soapmakers usually blend tallow with coconut oil, and this mixture is saponified. The resulting soap contains mainly the salts of palmitic, stearic, myristic, and oleic acids from the tallow, and the salts of lauric and myristic acids from the coconut oil. Coconut oil is added to produce a softer, more soluble soap.

Tallow $CH_3(CH_2)_{14}COOH$ $CH_3(CH_2)_{16}COOH$
Palmitic acid Stearic acid

$$CH_3(CH_2)_7CH=CH(CH_2)_7COOH$$
Oleic acid

Coconut oil $CH_3(CH_2)_{10}COOH$ $CH_3(CH_2)_{12}COOH$
Lauric acid Myristic acid

Depending on the source of the fat and the base used, you get soaps with different properties. Lard from hogs differs from tallow from cattle or sheep in that lard contains more oleic acid. Because the salt of a saturated long-chain acid makes a harder, more insoluble soap, the soap formed from tallow is less soluble than that formed from lard. Using different bases to saponify the fats also changes the properties of the soap. Sodium hydroxide produces a harder soap than that made from potassium hydroxide.

Soaps that we commonly use for bathing (toilet soaps) have been carefully washed free of any base remaining from the saponification. Floating soaps, such as Ivory soap, are produced by blowing air into the soap as it solidifies. Scouring soaps have added abrasives, such as fine sand or pumice.

Those of you who live in areas with hard water are probably familiar with a common problem with soaps. Soaps react with divalent cations to form insoluble precipitates we commonly refer to as "soap scum."

Hard water contains divalent metal cations, such as Fe^{2+}, Ca^{2+}, and Mg^{2+}.

$2\ RCOO^-\ Na^+\ +\ Mg^{2+}(aq)\ \rightarrow\ (RCOO^-)_2\ Mg^{2+}\ +\ 2\ Na^+(aq)$

Water-soluble Precipitate

An ion exchange column can be used to soften water. The column contains synthetic resin that holds sodium ions (or potassium ions). As hard water passes through the resin, the hard water ions are attracted to the beads and exchange themselves for the sodium or potassium ions. This process produces water that contains sodium or potassium ions instead of the divalent metal ions.

MATERIALS

Equipment
reflux apparatus
hotplate with sand bath
boiling stones
beakers
thermal gloves
watch glass
side-arm filter flask
Büchner funnel
10-mL graduated cylinder
test tubes

Chemicals
3 M NaOH in 50:50 water:ethanol
coconut oil or solid shortening
saturated NaCl solution
phenolphthalein solution
mineral oil
1% aqueous $CaCl_2$
1% aqueous $MgCl_2$
1% aqueous $FeCl_3$

CAUTION:
Use thermal gloves when handling hot glassware.

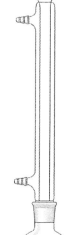

PROCEDURE

Part A: Synthesis of Soap

1. Assemble a reflux apparatus (see diagram in sidebar). The reflux apparatus consists of a round bottom flask and a reflux condenser. Be sure to use a ring stand and clamp to support the apparatus.

2. Place about 10 g of oil or shortening in a 100-mL round-bottom flask.

3. Add 30 mL of sodium hydroxide solution to the flask. Be sure to add 2–3 boiling stones. The boiling stones are required to prevent the reaction mixture from bumping. If no boiling stones are added or the mixture is heated to rapidly the mixture can bump all the way out of the top of the condenser!

4. Reflux the mixture for 35 minutes (hard boil).

5. Within about 20 minutes, enough soap should have formed such that you no longer see two layers in the reaction flask. Reflux for at least 10 more minutes, then turn off your burner and carefully pour the hot reaction mixture into a beaker containing 60 mL of saturated sodium chloride solution. At this point, the soap should precipitate out of solution.

6. Stir the mixture for several minutes. Using a spatula, break up any large lumps as completely as possible to permit maximum contact between the solid and wash solution.

7. Decant the wash solution from the soap, holding a watch glass over the mouth of the beaker to keep the soap from falling out. Repeat the washing with 30 mL more of saturated NaCl. The soap will have a granular texture.

8. Set up a filtration apparatus using a Büchner funnel (upper portion of figure in sidebar) and side-arm flask (lower portion). To set up a vacuum filtration, clamp the filter flask to a ring stand and use a black rubber hose to attach to the vacuum source. Place the vacuum adapter and then the funnel into the top of the flask. Do not forget to use filter paper in the funnel! Collect the wet

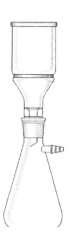

granules by vacuum filtration. Wash the granules with two portions (10–15mL) of water.

9. Use about 2/3 of the soap in Part C for making a bar of soap. Use the other 1/3 in Part B.

Part B: Testing Your Soap

Alkalinity

In this test, you will determine if you successfully washed out all the sodium hydroxide from your synthesized soap. The commercial soap, however, should have no residue of sodium hydroxide. Phenolphthalein is an acid/base indicator that is clear under acid conditions but pink in the presence of base.

1. Dissolve a pea-sized piece of soap in 5 mL of deionized water.
2. Add two to three drops of phenolphthalein indicator solution. Note the color.
3. Repeat this test using a commercial soap of known purity (e.g., Ivory). Note the color.

Emulsifying properties

Emulsifying agents have the ability to help two insoluble solutions mix together.

1. Prepare a soap solution by dissolving a small amount of the wet granular soap in 30 mL of deionized water. This solution will be used for testing emulsifying properties and behavior in hard water.
2. Place four drops of mineral oil into two separate test tubes.
3. Add 5 mL of deionized water to one test tube and 5 mL of the soap mixture to the other. Try to just get the soap solution, no soap granules.
4. Shake each tube vigorously for about 1 minute.
5. Observe the degree of oil emulsification in each tube as indicated by the presence of suds and the absence of oil droplets in the liquid or the absence of oil scum on the inside of the test tube. Describe the difference between the two tubes.

Behavior in hard water

1. Place 5 mL of soap solution in each of four test tubes.
2. Add 2 mL of 1% aqueous calcium chloride ($CaCl_2$) to the soap mixture in the first test tube. Add 2 mL of 1% aqueous magnesium chloride ($MgCl_2$) to the soap mixture in the second test tube. Add 2 mL of 1% aqueous ferric chloride ($FeCl_3$) to the soap mixture in the third test tube. In the fourth test tube, add regular tap water.
3. Mix each test tube and note whether a precipitate formed.
4. Add 4 drops of mineral oil to each tube and shake vigorously. Note how the emulsifying ability of the soap in these three test tubes compares with the emulsifying ability of soap in the absence of Ca^{2+}, Mg^{2+}, or Fe^{3+} ion.

Part C: Preparation of a Bar of Soap

1. Prepare a small paper boat from a piece of filter paper. Try to make a boat with bottom dimensions no more than 2.5 cm (≈1 in.) on each edge.

2. Weight the wet granular soap (2/3 of total) and place it in a 50-mL beaker. For each gram of soap, add 1 mL of deionized water. Slowly and carefully heat this mixture as you stir it. Keep the beaker from tipping by using crucible tongs. The next part of the operation is tricky. You want to bring the soap into a thick solution that is free of lumps, but DO NOT let the solution boil. If it works for you, you should eventually obtain a solution with a narrow upper band of froth.

3. Place the froth into your paper boat and let it cool. It will not take long for it to set hard enough to retain its shape. Let the product sit until the next lab period. The final bar may still be rather soft, but it should eventually become satisfactorily hard.

Name: _____

RESULTS

1. Describe the appearance of your product, soap.

2. Report the results of the alkalinity test.

Soap tested	Color of soap solution tested with phenolphthalein	Acid or base?

3. Describe the different between the two tubes in the emulsifying test.

4. Did you see a precipitate in any of the test tubes tested with metal cations? Describe any differences you see between the three test tubes, if any.

5. Describe the difference in the emulsifying ability of soap in the presence of the metal cations.

6. Compare the previous results with the test with tap water. Do you believe you have hard or soft water in your area? Justify your answer.

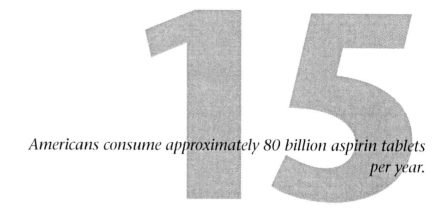

Americans consume approximately 80 billion aspirin tablets per year.

SYNTHESIS of ASPIRIN

Salicylic acid and its esters are found in several plants. The salicylic acid found in willow bark and leaves (*Salix* species) is the natural precursor for aspirin and has been used medicinally by native people all over the world. The entire willow plant contains pain-killing constituents (analgesics), with the highest concentrations found in the inner bark. Hippocrates (approximately 400 B.C.E.) recorded the use of willow bark and leaves to relieve pain and fever. It wasn't until the 1800s that the active component from willow, salicin, was isolated. Salicylic acid is derived from salicin but is converted to aspirin (acetylsalicylic acid) because salicylic acid is very hard on the stomach.

Willow bark tea is commonly used as a pain-killing agent.

The synthesis of aspirin from salicylic acid is an esterification reaction. Esters are formed from the acid-catalyzed reaction of an alcohol and a carboxylic acid. Salicylic acid (*o*-hydroxybenzoic acid) is a bifunctional compound. It is a phenol (an aromatic alcohol) and a carboxylic acid (R-COOH). Hence, it has the capacity of undergoing two different esterification reactions, acting as either the alcohol or the acid partner in the reaction. In the presence of acetic anhydride, acetylsalicylic acid (aspirin) is formed (reaction shown on next page). In this reaction, the salicyclic acid is reacting as an alcohol and the acetic anhydride is reacting as if it were a carboxylic acid.

In the reaction to prepare aspirin, the most likely impurity in the final product is salicylic acid itself. This impurity is generally the result of incomplete acetylation or hydrolysis of the product during the isolation steps. Salicylic acid, like most phenols, forms a highly colored complex with ferric chloride (Fe^{3+} ion). Aspirin, which has the hydroxyl group acetylated, will not give the color reaction. Thus, the presence of this impurity in the final product is easily detected.

An acetyl group

Salicylic acid   Acetic anhydride   Acetylsalicylic acid   acetic acid
                                              (aspirin)

Besides aspirin, acetic acid is a product of this reaction. This product is readily soluble in water and will be remove when the product is filtered. Dilute acetic acid is vinegar. Have you ever noticed that a bottle of aspirin tablets can have a distinct vinegar smell? As the aspirin tablets age the acetylsalicylic acid reacts with moisture and decomposes to acetic acid and salicylic acid. If your bottle of aspirin is old and has a strong vinegar smell you are more likely to experience the unwelcome side-effects seen with salicylic acid. Check the expiration date; you may want to throw it away. Many manufacturer now coat the aspirin tablets to prevent decomposition.

MATERIALS

Equipment
125-mL Erlenmeyer flask
400-mL beaker
heat source
thermometer
dispensing pipet
crystallizing dish
glass rod
10-mL graduated cylinder
100-mL graduated cylinder
Büchner funnel
filter flask
test tubes (3)

Chemicals
salicylic acid
acetic anhydride
conc. sulfuric acid
4% phenol solution
2.5% ferric chloride solution

CAUTION: Concentrated sulfuric acid is highly corrosive. You must handle it with great care.

PROCEDURE

Part A: Synthesis of Aspirin

1. Using a 400-mL beaker, set up a water bath and begin heating it with a heat source. You want the water to reach a temperature of about 75°C.

2. While the water is heating, weigh 2.0 g (0.015 mole) of salicylic acid crystals and place them in a 125-mL Erlenmeyer flask. Add 5 mL (0.05 mole) of acetic anhydride, followed by 1 to 2 drops of concentrated sulfuric acid and gently swirl the flask until the salicylic acid dissolves.

3. Heat the flask gently in the water bath for 10 to 15 minutes.

4. Allow the flask to cool to room temperature, during which time the acetylsalicylic acid should begin to crystallize from the reaction mixture. If it does not, scratch the inside walls of the flask with a glass rod.

5. Cool the mixture slightly in an ice bath until complete crystallization has occurred. If crystallization has failed to occur, you may also use a seed crystal (obtain from instructor) to create a nucleus for crystal formation.

6. After crystallization is complete, add 50 mL of water and cool the mixture in an ice bath. Usually, the product will appear as a solid mass when crystallization has become complete.

7. Collect the product by vacuum filtration. Set up a filtration apparatus using a Büchner funnel (upper portion of figure in sidebar) and side-arm flask (lower portion). To set up a vacuum filtration, clamp the filter flask to a ring stand and use a black rubber hose to attach to the vacuum source. Place the vacuum adapter and then the funnel into the top of the flask. Do not forget to use filter paper in the funnel! Cold water can be used to rinse the Erlenmeyer flask to collect all the crystals.

8. Rinse the crystals several times with small portions of cold water. Continue drawing air through the crystals for air drying.

9. Weigh the crude product, which may contain some unreacted salicylic acid.

Scratches on the inside of the flask create a rough surface upon which crystallization can occur.

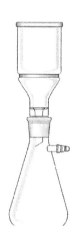

Part B: Testing Purity of the Product

1. To test the purity of your product, set up three test tubes as follows:

   Test tube 1   5 mL water + 10 drops 4% phenol

   Test tube 2   5 mL water + a few crystals of salicylic acid

   Test tube 3   5 mL water + a few crystals of product

2. Add 10 drops of 2.5% ferric chloride solution to each tube and note the color. Formation of an iron-phenol complex with Fe(III) gives a color ranging from red to violet, depending on the particular phenol. The presence of color in test tube 3 is an indication of an impurity.

Your instructor may ask you to save the product from this reaction to use in a later lab experiment, where you can examine the success of your synthesis of acetaminophen by using thin-layer chromatography (TLC).

Name: _____

RESULTS

Part A: Synthesis of Aspirin

1. Record the masses of salicylic acid used and of product formed:

Mass salicylic acid: _____

Mass aspirin: _____

2. Calculate the theoretical yield of acetylsalicylic acid. The **theoretical yield** is the potential mass of product. It is based on the mass of reactants and the stoichiometry of the reaction. The stoichiometry of the reaction is 1 mole of salicylic acid reacting with 1 mole of acetic anhydride to form 1 mole of aspirin.

   a. Calculate the moles of salicylic acid.

   $$\text{Moles salicylic acid} = \frac{\text{mass salicylic acid}}{\text{MW salicylic acid}}$$

   b. In this experiment, you used 0.05 moles of acetic anhydride. Which reactant is the limiting reagent, salicylic acid or acetic anhydride?

   The **limiting reagent** is the reactant that is completely consumed in a reaction.

   c. To determine the theoretical yield, you need to multiply the molecular weight of aspirin by the moles of limiting reagent.

   Theoretical yield = (moles limiting reagent) × (MW acetylsalicylic acid)

3. Determine the percent yield of product.

   $$\% \text{ yield} = \frac{\text{mass of product}}{\text{theoretical yield}} \times 100$$

Part B: Testing Purity of the Product

1. Record the results of the ferric chloride test.

Test tube	Color
1	
2	
3	

2. On the basis of the results of the ferric chloride test, comment on the purity of your product.

QUESTIONS

1. Why was sulfuric acid added to the reaction mixture?

2. Give some reasons why your yield was not 100%. If your yield was greater than 100%, give a potential cause.

3. As stated in the introduction, salicylic acid is a bifunctional compound. Draw and name the product of the esterification reaction of salicylic acid and methanol.

Acetaminophen is the mostly widely used pain reliever in the United States.

SYNTHESIS of ACETAMINOPHEN

Acetaminophen was first marketed in the United States in the early 1950s and has been available as a over-the-counter (OTC) pain reliever since the 1960s. Acetaminophen is the preferred pain reliever for people who will soon undergo surgery. Aspirin inhibits the aggregation of platelets and thus increases bleeding times, but acetaminophen has no impact on the platelet aggregation. Acetaminophen analgesic activity is believed to come from the drugs ability in increase the body's pain threshold. Acetaminophen is also an antipyretic and this is attributed to its action on the heat center in the hypothalmus. Acetaminophen is not considered a NSAID (non-steroidal anti-inflammatory drug) since it has no anti-inflammatory action.

Analgesic: pain reliever
Antipyretic: fever reducer

The synthesis of acetaminophen from *p*-aminophenol is the formation of a amide. Amides are formed from the reaction of an amine and a carboxylic acid. *p*-Aminophenol is a bifunctional compound. It is a phenol (an aromatic alcohol) and an amine ($R-NH_2$). Hence, it has the capacity of undergoing two different reactions. The reaction with acetic anhydride (acting as the carboxylic acid) can occur with either the alcohol or the amine group in the *p*-aminophenol. But luckily for those wanting the synthesized acetaminophen, the amine is more reactive. The reaction for the formation of acetaminophen is shown on the next page.

Besides acetaminophen, acetic acid is a product of this reaction. This product is readily soluble in water and will be removed when the product is filtered. Diluted acetic acid is vinegar; you will probably detect the smell of vinegar during the reaction.

An acetyl group

p-Aminophenol   Acetic anhydride   Acetaminophen   acetic acid

One of the ways to determine the purity and identity of a compound it to use its melting point. A pure compound will melt at a specific temperature, and the range of temperatures during the melting will be narrow, generally just a few degrees. For example, the melting point range of pure acetaminophen is 169–172°C. When impurities are present, the melting point of a compound broadens (the range can be 10°C or larger) and decreases. The impurity interferes with the crystal structure of the compound and causes the melting to take place at a lower temperature. As an optional exercise, you can take the melting point of your product and use the melting point to give you insight into the purity.

MATERIALS

Equipment
125-mL Erlenmeyer flask
400-mL beaker
heat source
thermometer
disposable pipet
glass rod
10-mL graduated cylinder
watch glass
Büchner funnel
filter flask
melting point apparatus

Chemicals
p-aminophenol
acetic anhydride

PROCEDURE

Part A: Synthesis of Acetaminophen

1. Using a 400-mL beaker, set up a water bath, and begin heating with a heat source. You will only need about 50 mL of water since only the bottom of the flask needs to be submerged in water. You want the water to reach a temperature of 75–85°C.

2. While the water is heating, weigh 1.0 g of *p*-aminophenol crystals and place them in a 125-mL Erlenmeyer flask. Use your 10-mL graduated cylinder to measure 3 mL of water and then add it to the flask to act as a solvent. Lastly, add 1.5 mL of acetic anhydride and gently swirl the flask. Do not worry if all the solid does not immediately dissolve—heating will help dissolve the crystals.

3. Heat the flask gently in the water bath for 15–20 minutes. Swirl the flask intermittently to help the crystals dissolve.

4. Once the reaction is complete, allow the flask to cool to room temperature. Crystals of acetaminophen should begin to form.

5. Cool the mixture slightly in an ice bath until complete crystallization has occurred. If crystallization has failed to occur, scratch the inside walls of the flask with a glass rod.

6. Collect the product by vacuum filtration. The vacuum apparatus consists of a Büchner funnel (upper portion of figure in sidebar) and side-arm flask (lower portion of figure in sidebar). To set up a vacuum filtration, clamp the filter flask to a ring stand and use a black rubber hose to attach to the vacuum source. Place the vacuum adapter and then the funnel into the top of the flask. Do not forget to use filter paper in the funnel! Break up the crystals so they will be easy to pour into the funnel. Cold water can be used to rinse the Erlenmeyer flask to collect all the crystals.

8. Rinse the crystals several times with small portions of cold water and continue to draw air through the crystals to aid in drying.

9. Weigh a watch glass and record the weight in the Results section.

10. Carefully use a spatula to move the crystals to the watch glass. Weigh the crystals and watch glass and record the weight in the Results section.

Scratches on the inside of the flask create a rough surface upon which crystallization can occur.

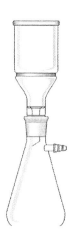

Part B: Determination of Purity of Product by Melting Point

1. To test the purity of your product, take the melting point of your product. The melting point of pure acetaminophen is 169–172°C. A pure, dry sample of acetaminophen should melt within 2–4°C of this range.

Your instructor may ask you to save the product from this reaction to use in a later lab experiment, where you can examine the success of your synthesis of acetaminophen by using thin-layer chromatography (TLC).

Name: _____

RESULTS

Part A: Synthesis of Acetaminophen

1. Record the masses of *p*-aminophenol used and of product formed:

   Mass *p*-aminophenol: _____

2. Determine the mass of acetaminophen formed:

   Mass of acetaminophen + watch glass: _____

   Mass of watch glass: (-) _____

   Mass of acetaminophen: _____

3. Calculate the theoretical yield of acetaminophen. The **theoretical yield** is the potential mass of product. It is based on the mass of reactants and the stoichiometry of the reaction. The stoichiometry of this reaction is 1 mole of *p*-aminophenol reacts with 1 mole of acetic anhydride to form 1 mole of acetaminophen.

   a. Calculate the moles of *p*-aminophenol.

   $$\text{Moles } p\text{-aminophenol} = \frac{\text{mass } p\text{-aminophenol}}{\text{MW } p\text{-aminophenol}}$$

   > The **limiting reagent** is the reactant that is completely consumed in a reaction.

   b. In this experiment, you used 0.016 moles of acetic anhydride. Which reactant is the limiting reagent, *p*-aminophenol or acetic anhydride?

   c. To determine the theoretical yield, you need to multiply the molecular weight of acetaminophen by the moles of limiting reagent.

   Theoretical yield = (moles limiting reagent) × (MW acetaminophen)

4. Determine the percent yield of product.

   $$\% \text{ yield} = \frac{\text{mass of product}}{\text{theoretical yield}} \times 100$$

Part B: Purity of the Acetaminophen

1. Record the melting point of the pure product.

2. On the basis of the melting point, comment on the purity of your product.

QUESTIONS

1. Give some reasons why your yield was not 100%. If your yield was greater than 100%, give a potential cause.

2. In the introduction, it was mentioned that acetic anhydride could react with the hydroxy group of *p*-aminophenol. Draw the structure of this compound.

3. Phenacetin, a derivative of acetaminophen, was introduced as a pain reliever in the 1890s long before the introduction of acetaminophen. Due to side effects, it is currently only used in special cases. Phenacetin differs from acetaminophen in that its hydroxy group has been modified into a methyl ether. Draw phenacetin.

Analgesic overdoses top the list of those reported by poison control centers.

DETECTION of ACETAMINOPHEN POISONING

Acetaminophen, or Tylenol®, structure shown below, is an effective analgesic and antipyretic agent. It has been used increasingly as a therapeutic agent because it lacks many of the undesirable side effects of aspirin. Acetaminophen is less irritating to the stomach and also lacks a side effect that makes aspirin unsuitable for children —the rare chance that a child taking aspirin for a flu or a cold will come down with the potentially fatal nervous system complication known as Reye's syndrome. This has led to an increase in the use of acetaminophen in pediatric products.

Analgesic: pain reliever
Antipyretic: fever reducer

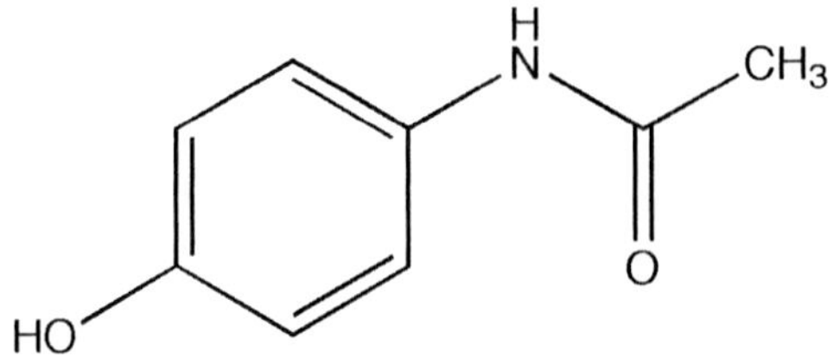

Acetaminophen, or
N-acetyl-*p*-aminophenol

Acetaminophen fatalities, which include both accidental and suicidal cases, dominate statistics gathered by poison control centers. This is due more to its availability than to its toxicity. Due to the over-the-counter nature of the drug, people often believe that it is safe at all doses or do not carefully monitor their intake of the drug. Initital symptoms of acetaminophen poisoning are nausea, vomiting, and abdominal pain. Eventually, acetaminophen overdoses can lead to irreversible liver damage and death. Acetaminophen levels greater than 200 µg/mL in serum at 4 hours after ingestion or 50 µg/mL at 12 hours are associated with severe liver damage. At these high levels, the drug exceeds the capacity of the liver to detoxify. This can be remedied with prompt detection and treatment. To this end, a simple chemical test has been developed that will not only detect acetaminophen in serum but will also give the doctor information about the amount of drug present.

You will not be using an actual serum sample but will be using a simulated serum sample.

In this experiment, you will determine the concentration of acetaminophen in an unknown "serum" sample. In order to assess the presence of the drug, we need to convert it to a form that can be easily assessed. For example, if we could convert acetaminophen into a colored product, we could measure its concentration by measuring the absorbance of the solution in a spectrophotometer. Beer's law states that the absorbance of a solution is proportional to the concentration of the solution. That is, the absorbance will vary linearly with concentration. If you plot the concentration of compound (in this case acetaminophen) *versus* absorbance, you can use this linear graph to determine an unknown concentration. This plot is called a **standard curve.**

Beer's law: $A = \varepsilon bc$
A = absorbance
ε = molar absorptivity
b = path length
c = concentration

In order to convert acetaminophen into a colored product, we first react the drug with nitrous acid.

In situ: formed at the site.

$$NaNO_2 + acid \rightarrow HO-N=O$$
Sodium nitrite   Nitrous acid

Nitrous acid is a weak, unstable acid, so it is necessary to form it *in situ* by the reaction of a strong acid with sodium nitrite. The reaction of acetaminophen with nitrous acid forms 3-nitroacetaminophen, which is then reacted with base to generate the yellow colored 3-nitroacetaminophen anion.

3-nitroacetaminophen

Base, OH⁻

3-nitroacetaminophen anion (yellow)

The yellow color of the 3-nitroacetaminophen anion allows us to determine the serum concentration of acetaminophen through the preparation of a standard curve.

To prepare the standard curve, you will make a serial dilution of known concentrations of acetaminophen. A serial dilution is a process where you use solution A to make a more dilute solution B, then use solution B to make an even more dilute solution C, and so forth. This is in contrast to starting from scratch with the solute for each solution. There is a benefit in making serial dilutions in that you only need measure the solute once and from thenceforth you use the initial concentration to make all subsequent concentrations. This is also the risk in making serial dilutions because, if you make a mistake in making the initial solution, then all subsequent solutions are incorrect. Don't worry the initial solution will be made for you, just carefully follow the instructions are given below for preparing the serial dilution. After preparing these samples, you will convert the acetaminophen to the colored 3-nitroaminophen anion. You will then determine the absorbance of each solution and construct a graph of absorbance (A) *versus* concentration (c). You will then use this curve to determine the concentration of an unknown serum sample.

MATERIALS

Equipment
test tubes

test-tube rack

10-mL graduated cylinder

large beaker

heat source

Spec 20 or equivalent

colorimeter tube

Chemicals
acetaminophen standard, 600 µg/mL

3% trichloroacetic acid

sodium nitrite solution, 0.07 M NaNO$_2$

6 M NaOH

serum unknowns

PROCEDURE

Part A: Preparing the Serial Dilution

1. Label four large test tubes 1A through 4A.
2. Add 3.0 mL of deionized water to test tube 2A and 4A and 4.5 mL of deionized water to test tube 3A.
3. Add 10.0 mL of acetaminophen standard (600 µg/mL) to test tube 1A.
4. Using your graduated cylinder, add 6 mL of the solution from test tube 1A to test tube 2A and mix the contents.
5. Add 4.5 mL of the solution from test tube 2A to test tube 3A and mix the contents. CAUTION: Be sure to rinse your graduated cylinder between each dilution.
6. Add 3.0 mL of the solution from test tube 3A to test tube 4A and mix the contents.

Part B: Reaction with Nitrous Acid

1. Label five new test tubes 1B to 5B. Starting with the most dilute solution, add 1.0 mL from each of the serial dilution test tubes (4A→1A) to the corresponding

> Starting with the most dilute solution will minimize any possible contamination of the samples.

new test tube (4B→1B). For test tube 5B, add 0.25 mL of the acetaminophen results serum "unknown."

2. To each test tube, add 6 mL of 3% trichloroacetic acid solution.

3. To each test tube, add 1.0 mL of sodium nitrite solution (0.07M NaNO$_2$) and mix well. Let these solutions heat in a water bath set for 37°C for 10 minutes. While your solutions are heating, turn on the Spec 20 and allow it to warm up.

4. After heating, add ten drops of 6M NaOH solution to each test tube. A yellow solution should form and persist.

Part C: Measurement of Absorbance

The absorbance of the samples will be measured at a wavelength of 430 nm. The reference blank will be a solution of 3% trichloroacetic acid. Be sure to use a colorimeter tube (Spec 20 test tube) for all absorbance measurements. Fill the colorimeter tube at least two-thirds full.

After the Spec 20 has warmed up for at least 10 minutes, the sequence for sample measurement is:

1. Turn on the instrument by turning the left knob on the front of the instrument clockwise.

2. Select the wavelength using the wavelength control knob on top of the instrument while watching the LED display of wavelengths.

3. Set the mode to TRANSMITTANCE (press the MODE select control until the transmittance LED on the right of the display is lit).

4. With the sample compartment empty and the cover closed, adjust the zero control so that the meter reads 0% T. The zero control is the same left front knob used to turn on the instrument.

5. Choose the mode that you require (absorbance) by pressing the MODE selector control until the absorbance LED (on the right of the display) is lit.

6. Add the reference blank (3% trichloroacetic acid solution) into the sample compartment and set 0.000A using the right knob on the front of the instrument.

7. Insert one of the samples into the sample compartment and read the measurement from the display in absorbance.

8. Recheck the 0% T after removing the sample. Periodically (at least twice during the readings) check the blank to be sure it reads 0.000A. If the instrument drifts from these readings, then readjust the appropriate knob in the front of the instrument and reread the samples affected by the instrument drift.

9. Turn off the instrument when finished.

Name: _____

DATA

Determination of the Concentrations of the Serial Dilutions

Use the formula $C_1V_1 = C_2V_2$ to determine the concentrations of the serial dilution. Test tube 1 contains a solution that has a concentration of 600 µg/mL. You took 6 mL (V_1) of 600 µg/mL (C_1) solution and put it into test tube 2. To this 6 mL, you added 3.0 mL of water for a final volume of 9.0 mL (V_2). What is the final concentration (C_2)?

$C_1V_1 = C_2V_2$ (600 µg/mL)(6 mL) = C_2 (9.0 mL)

3600 µg = C_2 (9.0 mL)

C_2 = 3600 µg/9.0 mL = 400 µg/mL

$C_1V_1 = C_2V_2$
C is concentration
V is volume

Fill in the rest of the table below. Remember, when calculating the concentration of test tube 3, that you are starting with the 400-µg/mL solution, not the 600-µg/mL solution.

Test tube #	Concentration	Absorbance
1B	600 µg/mL	
2B	400 µg/mL	
3B		
4B		
5B		

RESULTS

Standard Curve

Using the grid on the next page, plot the absorbance (A) versus concentration (c) of test tubes 1B to 4B. Also, include the zero concentration point of 0.000 absorbance. Draw the best-fit line through the points. Using this line, determine the concentration in your serum sample from its absorbance.

Unknown sample: _____

Concentration of unknown serum sample: _____

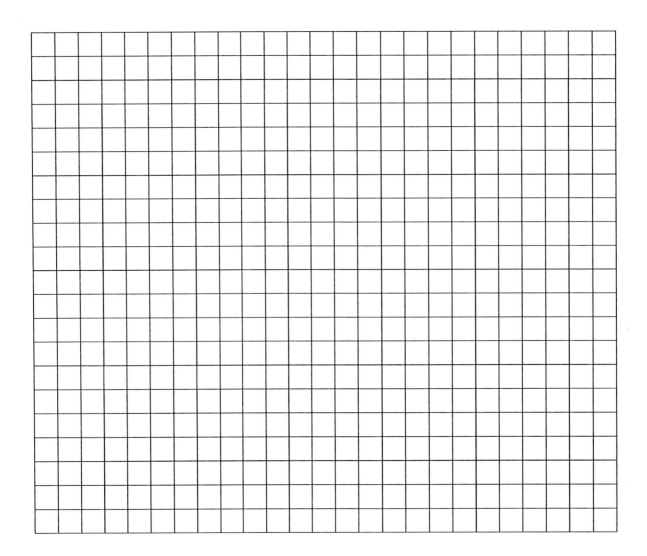

Assuming the serum sample was taken 4 hours after ingestion, did the person overdose on acetaminophen?

Many native peoples have used natural analgesics such as salicylic acid and its derivatives to relieve pain, reduce inflammation, and lower fevers.

THIN-LAYER CHROMATOGRAPHY of ANALGESICS

In this experiment, you will use a separation technique called thin-layer chromatography (TLC). TLC is often used in organic chemistry to monitor the progress of a reaction or to identify the components of a mixture. In TLC, a small drop of the mixture to be separated is applied near one end of a flexible plate coated with silica gel. The end of the plate is then dipped into a developing solvent, called the **mobile phase,** which flows up the plate by capillary action. As the developing solvent flows up the plate, it can carry along the components of the mixture. The rate at which a particular component moves depends on whether it tends to dissolve in the developing solvent or to remain adsorbed (or stuck) on the surface of the silica gel, called the **stationary phase.** A component that moves rapidly is spending more time dissolved in the developing solvent, whereas a component that moves slowly is spending more time adsorbed to the silica. The result is a separation of components of a mixture. A diagram of a typical chromatogram is shown below.

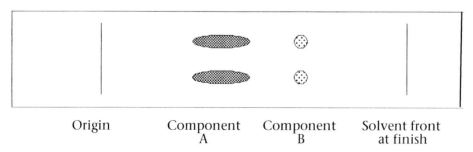

The results of a chromatographic separation are expressed in terms of R_f values. An R_f is the *relative* distance that a sample component has moved, that is, relative to the distance moved by the developing solvent. Using the chromatogram on the previous page as an example we can calculate the R_f's of the two components. Distance is measured from the point where the sample is applied (the origin) to the middle of each component spot. It is also necessary to measure the distance from the origin to the location of the solvent front. In the example, Component A traveled 4 cm and the developing solvent traveled 10 cm. The R_f of component A is calculated as follows:

$$R_f(A) = \frac{\text{distance A traveled}}{\text{distance solvent traveled}} = \frac{4 \text{ cm}}{10 \text{ cm}} = 0.4$$

R_f values determined under identical conditions are reasonably constant, so your calculated R_f values can be compared with another person's data or known values.

An **analgesic** is a compound capable of relieving pain.

In this experiment, you will use TLC to identify the compounds found in analgesic tablets. Depending on the brand of analgesic, the active ingredient is most likely aspirin, acetaminophen, or ibuprofen. In some cases, caffeine is added as a stimulant. You will use known samples of acetaminophen, acetylsalicylic acid (aspirin), caffeine, and ibuprofen and compare them by TLC with the mixtures found in the analgesic tablets. The structures of these reference compounds are shown below.

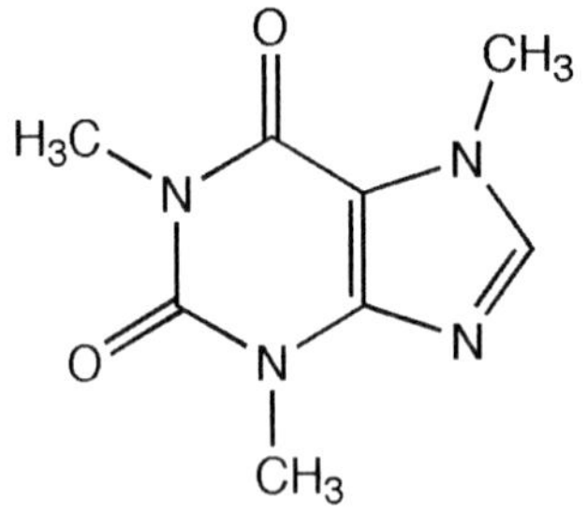

Caffeine

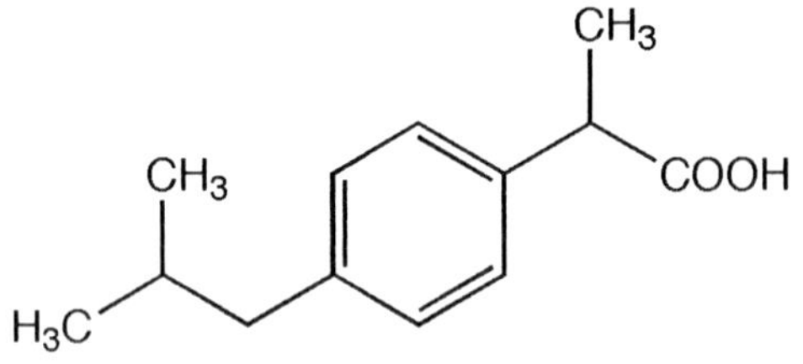

Ibuprofen

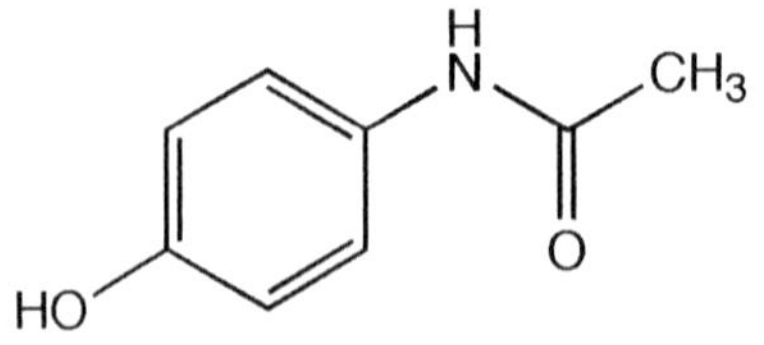

Acetaminophen

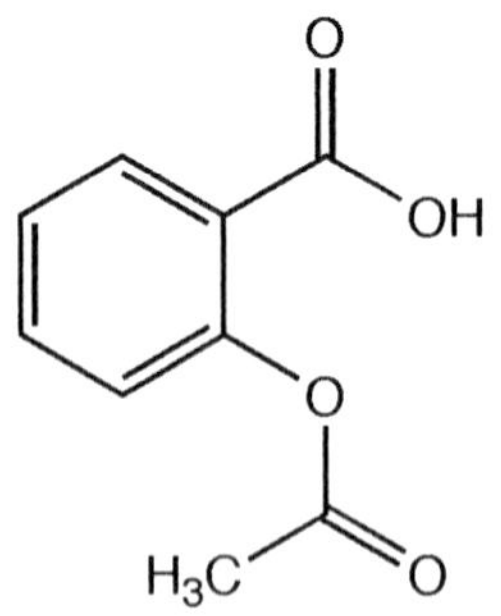

Aspirin

The method used to visualize the compounds will be illumination with ultraviolet (UV) light. The TLC plates are coated with a silica gel containing a fluorescent indicator. Under UV light, some of the spots will appear as dark areas on the plate, whereas others will fluoresce brightly. The differences in appearance under UV illumination, along with the distances the spots travel, will help to distinguish the substances from one another.

MATERIALS

Equipment
400-mL beaker
test tubes and rack
10-mL graduated cylinder
mortar and pestle
water bath
watch glass or Al foil
TLC plates
UV lamp
microcapillary tubes

Chemicals
5% acetaminophen (50:50 CH_2Cl_2:ethanol)
5% aspirin (50:50 CH_2Cl_2:ethanol)
5% caffeine (50:50 CH_2Cl_2:ethanol)
5% ibuprofen (50:50 CH_2Cl_2:ethanol)
assortment of analgesics
methylene chloride (CH_2Cl_2)
0.5% acetic acid in ethyl acetate
ethyl acetate
ethanol (CH_3CH_2OH)

PROCEDURE

1. Obtain a TLC plate; handle the plate carefully by the edges and do not bend. Using a lead pencil (not a pen), lightly draw a line across the plate (about 1 cm from the bottom). Using a centimeter ruler, lightly mark off four intervals equally spaced on the line. These are the points at which the samples will be spotted. You will spot each reference compound: acetaminophen, aspirin, caffeine, and ibuprofen.

2. When the plate has been spotted and it is dry, it is ready to be placed in a development chamber. For a development chamber, you will use a 400-mL beaker lined with a piece of filter paper and covered with a watch glass or aluminum foil. The purpose of the filter paper is to provide an atmosphere in the developing chamber that is saturated with solvent.

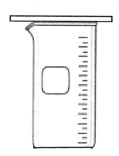

Development chamber

3. When the development chamber has been prepared, obtain a small amount of the development solvent (0.5% glacial acetic acid in ethyl acetate). Fill the chamber with the development solvent to a depth of about 1/4 inch. Be sure that the filter paper liner is saturated with the solvent. The solvent level must not be above the spots on the plate or the samples will dissolve off the plate into the reservoir instead of developing. Place the spotted plate in the chamber and allow the plate to develop.

4. When the solvent has risen to a level about 1 cm from the top of the plate, remove the plate from the chamber and, using a lead pencil, mark the position of the solvent front. Let the plate dry. When the plate is dry, observe it under a short-wavelength UV lamp. Lightly outline all the observed spots with a pencil.

5. While you are running the first TLC plate your will have time to prepare a second TLC plate with the different brands of analgesics. But be careful while you are making the second plate that you do not let the solvent run off the top of the developing TLC plate or you will have to do the first plate over.

6. Obtain half a tablet of each of the two analgesics to be analyzed. Crush the tablet by using a mortar and pestle—or you can crush it on a piece of filter paper with a spatula. Transfer each crushed half-tablet to a small, labeled test tube.

7. Using a graduated cylinder, mix together 10 mL of absolute ethanol and 10 mL of methylene chloride. Mix the solution well. Add 5 mL of this solvent to each of the test tubes and then heat each of them gently for a few minutes in a water bath. Not all the tablet will dissolve because the analgesics usually contain an insoluble binder and/or buffering agents and coatings.

8. Obtain a second TLC plate; again handle the plate carefully by the edges and do not bend. Using a lead pencil (not a pen), lightly draw a line across the plate (about 1 cm from the bottom). Using a centimeter ruler, lightly mark off four intervals equally spaced on the line. On this plate you will the two brands of analgesics, plus two of the reference compounds you believe to be in your samples.

9. After heating the samples, allow them to settle and then spot the clear liquid extracts on the plate. To spot the plate, dip a capillary tube into the extract to draw up some liquid into the tube. Lightly touch the capillary tube to the plate. Allow just a small spot of liquid onto the plate. Allow the spot to dry and spot again to get a more concentrated sample on the plate

10. When the plate has been spotted and it is dry, it is ready to be placed in a development chamber.

11. When the solvent has risen to a level about 1 cm from the top of the plate, remove the plate from the chamber and, using a lead pencil, mark the position of the solvent front. Let the plate dry. When the plate is dry, observe it under a short-wavelength UV lamp. Lightly outline all the observed spots with a pencil.

Name: _____

RESULTS

1. Name of tested brands of analgesics: a. _____

   b. _____

2. Draw a diagram of the chromatogram of the reference compounds.

   [ ]

3. Record any observations about the appearance of the spots under UV illumination.

4. Report the R_f value for the reference compounds.

Compound	R_f value
Acetaminophen	_____
Aspirin	_____
Caffeine	_____
Ibuprofen	_____

5. Draw a diagram of the chromatogram of the analgesics with reference compounds.

   [ ]

6. Calculate the R_f values for the components of your tested analgesic tablets and reference compounds. Use the R_f values and observations to identify all the components. Be sure to compare the R_f's of your tablets to all the reference compounds.

                                  R_f values   Identity of components

Tablet 1:

Tablet 2:

Reference 1:

Reference 2:

QUESTIONS

1. In examining your TLC of the analgesic tablets, did you see components other than the reference compounds? Explain.

2. In the introduction, we stated that TLC could be used to follow the progress of a reaction. Speculate on how this can be accomplished.

You will compare the specificity of the enzymes lactase and sucrase.

ENZYME SPECIFICITY of SUCROSE and LACTOSE HYDROLYSIS

Carbohydrates, also known as sugars or saccharides, are one of the most important classes of compounds encountered in biochemistry. These compounds function as a source of energy for living organisms as well as providing structural material for cells. Carbohydrates are complex molecules containing both alcohol and carbonyl functional groups.

Carbohydrates can exist as single sugar units called **monosaccharides** or as polymers called **polysaccharides**. An example of a monosaccharide, glucose, is shown below.

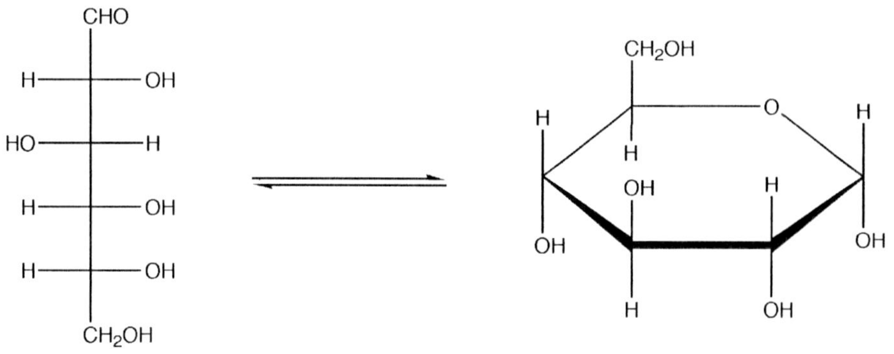

D-glucose (open chain)   ⇌   α-D-glucose (cylic form)

Because glucose has both alcohol and carbonyl functionalities, it is able to form a ring; in the above case the glucose is in the pyranose form (6-membered ring). You will also notice that this is the α anomer; that is, the hydroxyl group on carbon 1 is down. When monosaccharides are in solution an equilibrium exists between the open chain form and the two possible anomers, α and β.

When two monosaccharides react with each other they form a disaccharide. The reaction of one hydroxyl on one sugar with a hydroxyl on another forms a glycosidic linkage. To identify the disaccharide you need to know the sugars involved, the carbon atoms that contain the hydroxyl groups that join the two sugars, and the stereochemisry of the anomeric carbon. The structure below is for lactose.

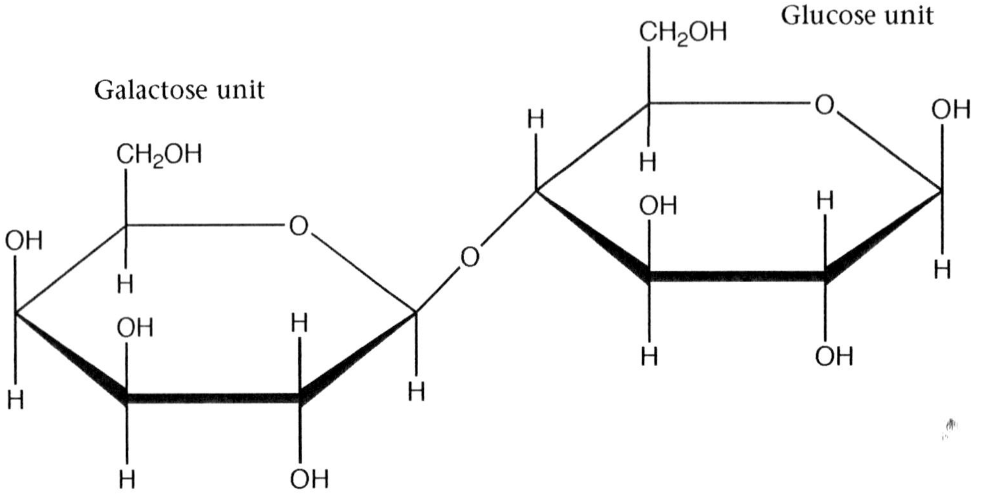

β-D-lactose

Lactose, which is a disaccharide, is composed of a galactose and glucose unit. In lactose, a glycosidic bond is formed with the hydroxyl group on carbon 1 of the β anomer of galactose. The hydroxy group on glucose that participates in the glycosidic linkage is on carbon 4. Therefore, the glycosodic linkage in lactose is referred to as an β(1→4) linkage.

The enzyme that cleaves the β(1→4) linkage in lactose is called lactase. People who are lactose intolerant either lack or have reduced amounts of the enzyme lactase. When people with lactose intolerance ingest milk products they suffer from gas, bloating, stomach cramps, and possibly diarhea. Lactase is commercially available for people with lactose intolerance. Lactase can be added to milk and it will cleave the lactose into galactose and glucose, which does not cause the symptoms seen with lactose. Lactase is also available in tablet form and can be taken before ingesting milk products.

Sucose is another dissacharide, only it contains the two monosaccharides glucose and fructose. Sucrose, is what you know as table sugar. The fructose and glucose units are joined by a glycosidic linkage between the anomeric carbons on both sugars. Notice in the structure given for sucrose the stereochemistry on the anomeric carbon of glucose is α, whereas the stereochemistry on the anomeric carbon of fructose is β. This linkage is referred to as an α1↔2β linkage, an a linkage from carbon 1 of glucose to a β hydroxyl on carbon 2 for fructose.

As stated earlier, sugars are used by the body as a source of energy. To get energy from table sugar (sucrose), your digestive system must first break it into the monosaccharide units glucose and fructose. In the case of sucrose the enzyme we use is sucrase, sometimes referred to as invertase because the product of the reaction is called invert sugar.

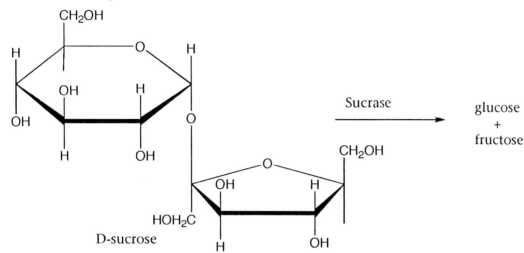

The hydrolysis of sucrose

The enzyme sucrase is often called invertase because the product of the hydrolysis of sucrose is invert sugar.

Invert sugar is a 50:50 mixture of glucose and fructose. Invert sugar is sweeter than table sugar. It is the form found in honey.

Lactase and sucrase are enzymes. Enzymes are protein catalysts that accelerate the rate of a reaction. Enzymes are unique among catalysis in that they have great catalytic power as well as specificity. The degree of specificity varies among enzymes. In this experiment you will examine the specificity of the two enzymes lactase and sucrase. Both enzymes perform hydrolysis reactions. They cleave the disaccharides into their component monosaccharides. You will determine if lactase is specific for lactose or will it also hydrolyze sucrose, and *vice versa*. You will use glucose test strips to look for the release of glucose upon the hydolysis of the two disaccharides as the indicator for cleavage of the glycosidic linkage.

MATERIALS

Equipment
small test tubes (7)

test-tube rack

10-mL graduated cylinder

250-mL beaker

thermometer

glass stirring rod

spatula

Chemicals
sucrose

lactose

glucose

aq. sucrase (40 mg/100 mL)

aq. lactase (1 tablet/200 mL)

glucose test strips

PROCEDURE

1. Add 0.3 g of sucrose to three different test tubes and label the tubes 1 to 3.
2. To test tube 1, add 5 mL of distilled water (this will be a control). To test tube 2, add 5 mL of aq. sucrase. To test tube 3, add 5 mL of aq. lactase. Shake the tubes to mix. Do not worry if all the sugar does not dissolve immediately; the sugar will dissolve upon heating.

Before pouring out the enzyme solution, shake the bottle to suspend any particulates.

3. Add 0.3 g of lactose to three different test tubes and label the tubes 4 to 6.

4. To test tube 4, add 5 mL of distilled water (this will be a control). To test tube 5, add 5 mL of aq. sucrase. To test tube 6, add 5 mL of aq. lactase. Shake the tubes to mix. Again, do not worry if all the sugar does not dissolve; the sugar will dissolve upon heating.

5. Place the six test tubes in a warm water bath at 35–40°C for 40 minutes. Shake the tubes periodically to help the sugars dissolve.

6. While test tubes are incubating, prepare an additional test tube with a small amount of glucose (pea size on tip of spatula) and 1 mL of water. You will use this to observe a positive test for glucose. Mix the glucose in the water and then dip the glucose test strip into the solution. The dipstick will come with a color chart that will indicate the approximate glucose concentration. Wait 30–60 seconds and note the color and concentration of the test in the Data section.

7. When the incubation of test tubes 1 to 6 is completed, allow the solutions to come to room temperature. Test each solution with a new dipstick. For each test tube, note the color and concentration in the Data section.

The results of the dipstick test will vary with temperature. Make sure all solutions have reached room temperature before testing.

Name: _____

RESULTS

Results of test of glucose: Color: _____

   Concentration: _____

Test tube (sample)	Result (color)	Concentration
1 (sucrose control)		
2 (sucrose + sucrase)		
3 (sucrose + lactase)		
4 (lactose control)		
5 (lactose + sucrase)		
6 (lactose + lactase)		

QUESTIONS

1. What are the product of hydrolysis of lactose and sucrose?

2. What result indicates hydrolysis of either lactose or sucrose?

3. Under what conditions did you see hydrolysis of sucrose?

4. Under what conditions did you see hydrolysis of lactose?

5. What does this experiment tell you about enzyme specificity?

6. What unit is used for the concentration of glucose? How does this unit differ from molarity?

The hydrolysis of sucrose can be accomplished with an enzyme or with acid.

HYDROLYSIS of SUCROSE

Sucrose is a disaccharide formed from the two monosaccharides glucose and fructose. Sucrose is also known as common table sugar. Sugars belong to a important class of biomolecules called carbohydrates. Carbohydrates function as a source of energy for living organisms as well as providing structural material for cells.

A **disaccharide** is a molecule composed of two monosaccharides.

In solution, monosaccharides exist at equilibrium between the open chain with a free aldehyde or ketone and a 6- or 5-membered ring form. Because both forms exist in solution, the chemistry of these sugars is similar to that of an aldehyde in the case of glucose and a ketone in the case of fructose. The aldehyde form of glucose

α-D-glucose (hemiacetal form) D-glucose (aldehyde form)

a-D-fructose (hemiacetal form) D-glucose (ketone form)

is easily oxidized. Carbohydrates that can be oxidized are known as **reducing sugars.** When urine is tested for the presence of glucose the test is looking at the ability of glucose to be oxidized—it is testing for the presence of a reducing sugar.

In a redox reaction, the compound being oxidized is the **reducing agent.**

As stated earlier, sucrose is a disaccharide composed of a fructose unit and a glucose unit. The fructose and glucose units are joined by a linkage between both the carbonyl carbons; this linkage is stable and the ring does not open. Because the ring does not exist in equilibrium with the open chain form, the aldehyde and ketone forms are unavailable. Because sucrose is not in equilibrium with an aldehyde form, it cannot be oxidized and is considered a **non-reducing sugar**.

The enzyme sucrase is often called invertase because the product of the hydrolysis of sucrose is invert sugar.

Invert sugar is a 50:50 mixture of glucose and fructose. Invert sugar is sweeter than table sugar. It is the form found in honey.

D-sucrose $\xrightarrow{\text{Sucrase}}$ glucose + fructose

To get energy we need from sucrose, our digestive system must first break sucrose into the monosaccharides units glucose and fructose. In the laboratory, this hydrolysis of the linkage between glucose and fructose is normally done by heating sucrose in a strong acid solution, for example, 1M HCl. However, most of the human body cannot tolerate such acidic conditions. Fortunately, living organisms can hydrolyze carbohydrates to gain glucose by using enzymes. **Enzymes** are protein catalysts that accelerate reactions. An enzyme that can hydrolyze sucrose to glucose and fructose at pH 7 is sucrase.

In this experiment, you will explore the hydrolysis of sucrose. You will hydrolyze sucrose into fructose and glucose by the two different methods discussed above. One solution of sucrose will be heated in $1M$ HCl and the other solution will be incubated with the enzyme sucrase. To see whether your hydrolyses were successful, you will test for the presence of reducing sugars using Benedict's reagent. In the Benedict's test, copper (II) is reduced to copper (I) by reducing sugars, a reaction that changed the color of the solution from light blue to colors ranging from dark green to brick red. While the solutions are incubating, you will use Benedict's reagent to test sucrose and glucose to see whether they are reducing sugars.

MATERIALS

Equipment
small test tubes (5)
test-tube rack
10-mL graduated cylinder
250-mL beaker
thermometer
glass stirring rod
spatula
pH paper

Chemicals
sucrose
glucose
$1M$ HCl
aq. sucrase/invertase (40 mg/100 mL)
$3M$ NaOH
Benedict's reagent

PROCEDURE

1. Add 0.3 g of sucrose to three different test tubes and label the tubes 1 to 3.
2. To test tube 1, add 5 mL of distilled water (this will be a control). To test tube 2, add 5 mL of $1M$ HCl. To test tube 3, add 5 mL of aq. sucrase
3. Place the three test tubes in a water bath and heat at 37°C for 50 minutes. Shake test tubes 2 and 3 periodically.
4. While test tubes 1 to 3 are incubating, prepare two additional test tubes labeled 4 to 5. To test tube 4, add 0.3 g of sucrose to 5 mL of distilled water. In test tube 5, add 0.3 g of glucose to 5 mL of distilled water.
5. When the incubation of test tubes 1 to 3 is completed, add 1 mL of $3M$ NaOH to test tube 2 to neutralize the acid. Check the pH with pH paper by dipping a glass rod into the solution and then touching it to the paper (do not insert the paper into the test tube). If the solution is still acidic, add base dropwise to the solution until the pH paper indicates the solution is no longer acidic.
6. Perform the Benedict's test on all five samples. For the Benedict's test add 5 mL of Benedict's reagent to the solutions in each test tube (1 to 5).
7. Heat these mixtures on a hot water bath for 10 minutes. A reducing sugar will produce a red, green or yellow precipitate; this is considered a positive test.

Before pouring out the enzyme solution shake the bottle to suspend any particulates.

Name: _____

RESULTS

Test tube (sample)	Result (color)	Positive/negative
1 (blank)		
2 (sucrose + HCl)		
3 (sucrose + sucrase)		
4 (sucrose)		
5 (glucose)		

Did you see hydrolysis of sucrose under both conditions? Do you see any difference between the two samples?

QUESTIONS

1. Honey is also known as invert sugar. In honey, the sucrose has been hydrolyzed into fructose and glucose. Is honey a reducing or non-reducing sugar? Explain.

2. Starch is a polysaccharide consisting of hundreds of units of glucose. Although glucose is a reducing sugar, starch is not. Explain this phenomenon.

Casein is the major protein in milk.

ISOLATION of CASEIN from MILK

As a source of nutrients there are few foods that can compete with milk. Milk contains protein, carbohydrates, fats, vitamins, and minerals. The major protein in milk is casein. Casein has a negative change at the pH of milk and is associated with calcium cations. Casein is considered a complete protein. A complete protein is a protein that contains all the essential amino acids. The other major protein in milk is bovine serum albumin. The properties of these two proteins are sufficiently different to allow the isolation of casein from milk without the isolation of bovine serum albumin. To isolate casein, acid will be added to the solution and the change in pH will change the protein from a soluble protein to an insoluble protein, which will precipitate out of solution.

Essential amino acids are the amino acids that we are unable to synthesize in sufficient quantitites for normal growth.

While proteins are the major component of milk another significant component is lactose. Lactose is a disaccharide composed of glucose and galactose. Some people are unable to cleave lactose into its two monosaccharide components due to the lack of sufficient quantities of the enzyme lactase. People who are deficient in lactase are lactose intolerant and need to avoid milk or take lactase tablets when they eat milk products.

In the nursery rhyme, little Miss Muffet sat eating her "curds and whey." When you precipitate the casein you will be making the curds, and the liquid left behind is whey.

Milk is also a great source of vitamins. Milk contains the fat-soluble vitamins A, D, E, and somtimes K. These vitamins are mainly found associated with the milk fat. The amount of these vitamins in milk are influenced by the diet of the mammal that produces the milk. Milk is often fortified with vitamin D and sometimes A to provide a good source for these vitamins. Milk also contains water-soluble B vitamins.

Milk fat is the final major component of milk. Cow's milk is approximately 4% fat, a third of which is unsaturated fat. Milk fat is unusual in that it contain triglycerides with relatively short fatty acid chains (C4-C10). If milk is homogenized the fat will not separate from the other components.

In this experiment you will isolate casein by the selective precipitation of this protein. What will be left behind in the supernatant of this precipitation will be the protein albumin and the carbohydrate lactose. You will test the supernatent for the presence of these components.

MATERIALS

Equipment

two 100-mL beakers

hotplate

stirring rod

rubber policeman

thermometer

Büchner funnel

filter flask

watch glass

filter paper

six test tubes

Chemicals

powdered milk

dilute aq. acetic acid (10%)

Bradford's reagent

Benedict's reagent

aq. bovine serum albumin (1 mg/mL)

lactose

PROCEDURE

Part A: Isolation of Casein

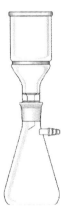

1. Put approximately 3 grams of powdered milk into a 100-mL beaker. Record the exact mass in the Data section. Add 10 mL of water to the beaker.

2. Place the beaker on a hot plate and heat to between 40 and 45°C.

3. Once the solution reaches temperature add acetic acid dropwise. You will notice that the milk will begin to curdle. This is the precipitation of the casein. It will look similar to cottage cheese. As the curds form, push them aside with the rubber policeman. Add more acetic acid; if more casein precipitates continue adding acetic acid until casein no longer precipitates. This step should require only a couple of mL's of acetic acid.

4. Set up a filtration apparatus using a Büchner funnel (upper portion of figure in sidebar) and side-arm flask (lower portion). To set up a vacuum filtration, clamp the filter flask to a ring stand and use a black rubber hose to attach to the vacuum source. Place the vacuum adapter and then the funnel into the top of the flask. Do not forget to use filter paper in the funnel!

5. Pour the casein into the Büchner funnel and allow air to flow through until you no longer see any liquid flowing into the filter flask. Shut off the vacuum and remove the casein from the filter and place on a piece of paper towel. Fold the towel over the casein and press to remove additional liquid.

6. Weigh a watch glass and record the mass in the Data section. Transfer the casein to the watch glass and let it dry. Once the casein is dry, weigh again and record the mass in the Data section.

7. Save the filtrate because you will test it for the presence of protein and sugar.

Part B: Test for the Presence of Lactose

The test you will use to look for the presence of lactose is the Benedict's test. Benedict's test looks for the presence of reducing sugars, that is, sugars that have an aldehyde that can be reduced. Lactose is a reducing sugar. While other sugars are reducing sugars, lactose is, by far, the sugar that is most likely to be found in milk. Benedict's solution is a clear aqua color. In the presence of a reducing sugar, Benedict's reagent will form a precipitate that ranges in color from a mustard yellow to brick red color.

1. Place 50 mL of water in a beaker and start heating on a hotplate. You want the water bath to be close to the boiling point for the Benedict's test.

2. Set up three test tubes for the Benedict's test. Add 5 mL of Benedict's reagent to each test tube.

   a) Add nothing to the first test tube; it will serve as a color control for a negative Benedict's test.

   b) Add 0.1 g of lactose to the second test tube This will serve as a positive control to allow you to see the color of a positive Benedict's test.

   c) Add 1 mL of the filtrate to the third test tube.

3. Place all three test tubes in the water bath and heat for about 5 minutes. Note any color change you see in the Data section.

Part C: Test for the Presence of Protein

The protein test will use Bradford's reagent, which changes from a light shade of blue to a darker shade of blue in the presence of protein.

1. Set up three test tubes for the protein test. Add 3 mL of water to each test tube.

   a) Add nothing to the first test tube; it will contain only water to serve as a color control for a negative protein test.

   b) Add 10 drops of bovine serum albumin (BSA) to the second test tube. This will serve as a positive control to allow you to see the color of a positive protein test.

   c) Add 1 mL of the filtrate to the third test tube.

2. Add 1 mL of Bradford's reagent to each test tube.

3. Allow the test tubes to sit for several minutes. Compare your filtrate with the positive and negative protein test and note the colors you see in each test tube in the Data section.

Name: _____

RESULTS

Part A: Isolation of Casein

Mass of powdered milk: _____

Mass of casein isolated: _____

    Mass of watch glass + casein   _____

    Mass of watch glass   (-)   _____

    Mass of casein   (=)   _____

Calculate the percent of milk that is casein:

    Mass of casein/mass of powdered milk × 100% = % casein

Part B: Test for the Presence of Lactose

	Color	Positive/Negative
Test tube 1 (water)	_____	_____
Test tube 2 (lactose)	_____	_____
Test tube 3 (filtrate)	_____	_____

Part C: Test for the Presence of Protein

	Color	Positive/Negative
Test tube 1 (water)	_____	_____
Test tube 2 (BSA)	_____	_____
Test tube 3 (filtrate)	_____	_____

QUESTIONS

1. a) If, as stated in the introduction to this experiment, casein has a negative charge at the normal pH of milk, what happened to the casein to cause the precipitation from solution during this experiment?

   b) Would you expect the calcium cations to still be associated with the casein when it precipitates?

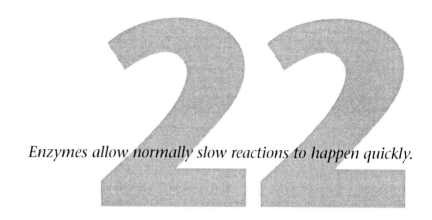

Enzymes allow normally slow reactions to happen quickly.

ANALYSIS of WHEAT GERM ACID PHOSPHATASE

An **enzyme** is a protein catalyst. Many reactions that take place in the human body are very slow. For example, you can keep a mixture of table sugar (sucrose) and water for a reasonably long period. Yet, when you eat table sugar, it is easily metabolized into its components, glucose and fructose. It is the presence of enzymes in your body that allows you to carry out reactions rapidly that normally take minutes, hours, or perhaps days. The simplest way to represent an enzyme-catalyzed reaction would be as follows:

A **catalyst** is a substance that increases the rate of a chemical reaction but is not itself changed in the process.

$$\text{Substrate (S)} \xrightarrow{\text{enzyme}} \text{product (P)}$$

The velocity or rate of the reaction can be either determined by measuring the decrease in substrate concentration, $-\Delta[S]$, with time, Δt, or, more commonly, by measuring the rate of appearance of product, $\Delta[P]$, with time:

$$\text{Rate} = -\Delta[S]/\Delta t = \Delta[P]/\Delta t$$

The rate of a reaction is an important property of an enzyme. Enzymes can increase the rate of a chemical reaction 10^3- to 10^{10}-fold. Factors that can affect the rate of an enzyme-catalyzed reaction are enzyme concentration, temperature, and pH. Temperature and pH will affect the enzyme itself; each enzyme has an optimal pH and temperature range. At temperatures or pH values outside the optimum, the enzyme can be denatured and inactivated. Therefore, it is important that these two factors be held constant during the measurement of the rate of the enzyme-catalyzed reaction. The rate of an enzyme-catalyzed reaction will be proportional to the concentration of substrate, up to a point. At high substrate concentrations, the enzyme will be saturated – that is, it is going as fast as it possibly can. At this point, added substrate will not increase the rate of reaction.

To determine the rate of an enzyme-catalyzed reaction, one uses a fixed concentration of enzyme and substrate. During the early part of the reaction, the amount of product formed increases linearly with time. However, in the latter part of the reaction, the rate of product appearance diminishes to a point where product is no longer formed. This is illustrated graphically below. There are a number of reasons for the decline in reaction rate with time. These include the depletion of substrate or the breakdown (denaturation) of the enzyme. Thus, the rate or velocity of the reaction is determined during the early time interval when the amount of product is increasing linearly with time. The rate is determined from the slope of a straight line tangent to the beginning of the curve. This rate is called the initial velocity of the reaction (v_0). The v_0 in the illustration below is approximately 10 nanomoles (nmol) of product formed per minute.

Wheat germ acid phosphatase catalyzes the hydrolysis of phosphate groups from macromolecules that are stored in the wheat seed. The growing wheat embryo uses the freed phosphate in germination and growth. In this experiment, you will measure the velocity of the reaction catalyzed by purified acid phosphatase. Nitrophenyl phosphate, a colorless compound, will be used as a substrate in the experiment. The hydrolysis products are nitrophenol and phosphate (see below).

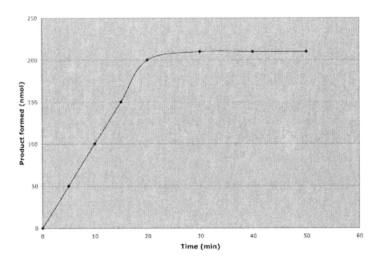

Nitrophenyl phosphate   Nitrophenol

Under alkaline conditions, nitrophenol is converted to a nitrophenoxide ion, which is yellow. Using a spectrophotometer, the concentration of product (nitrophenol) can be measured by the increase in absorbance due to the yellow color.

Nitrophenol (colorless) + OH⁻ → Nitrophenoxide (yellow) + H_2O

The addition of base at the end of the reaction serves two purposes. First, it converts the product into a colored substance that we can use to measure concentration. Second, it stops the reaction by denaturing the enzyme. In this way, we can determine the concentration of product formed at specific time intervals by simply adding base.

There are two parts to this experiment. First, you will use acid phosphatase to catalyze the hydrolysis of nitrophenyl phosphate. The reaction will be stopped by the addition of base at various time intervals. The concentration of nitrophenol can be determined by measuring the absorbance and extrapolating the concentration from the standard curve.

In the second part, you will need to make a standard curve. You will prepare a series of test tubes containing increasing concentrations of nitrophenol. You will measure the absorbance of the different concentrations under alkaline conditions. The absorbance will vary linearly with concentration. If you plot the concentrations of nitrophenol *versus* absorbance, you can use this curve to determine an unknown concentration.

In this experiment you will determine the rate of the reaction at the beginning of the reaction. This is the most accurate part of the curve because little product has been formed and substrate concentration is not depleted. You may not see the flattening of the curve seen on the illustration graph because your time is limited.

MATERIALS

Equipment
stopwatch
test tubes and rack
10-mL graduated cylinder
small erlenmeyer
Spec 20 or equivalent
1-mL transfer pipet
colorimeter tubes

Chemicals
1 mM nitrophenylphosphate in 0.9M citrate buffer, pH 4.8
0.1 M aqueous KOH
3.2 mM nitrophenol standard
wheat germ acid phosphatase

PROCEDURE

Part A: Enzyme Assay

1. Prepare a rack of 10 small test tubes that will fit in the spectrophotometer. Using a waterproof pen, number the tubes from 1 to 10.
2. Use a 10-mL graduated cylinder to place 7 mL of 0.1 M KOH into each of the 10 tubes.
3. Obtain approximately 15 mL of phosphatase substrate (nitrophenyl phosphate solution) in a small beaker. Using a transfer pipet, place 1 mL of substrate solution in test tube 1. This solution will be used to determine the zero-time value of the reaction.
4. Transfer 10 mL of the phosphatase substrate from the beaker into a small erlenmyer flask.
5. Using a 1-mL transfer pipet, place 0.5 mL (500 µL) of wheat germ acid phosphatase into the Erlenmeyer flask. Swirl the flask to mix. Begin timing!
6. At the times indicated below, remove 1 mL of the solution from the Erlenmeyer flask. Place the solution into the tube corresponding to the time of withdrawal.

Time interval	Test tube
0 minutes	1
1 minute	2
2 minutes	3
3 minutes	4
4 minutes	5
5 minutes	6
7.5 minutes	7
10 minutes	8
15 minutes	9
20 minutes	10

Part B: Preparation of the Nitrophenol Standards, a Serial Dilution

1. Number a set of test tubes from 1 to 6.
2. Add 2 mL of water to each test tube.
3. Add 2 mL of nitrophenol standard (3.2 mM) to test tube 6 and mix the contents.
4. Using your graduated cylinder, transfer 2 mL of the solution from test tube 6 to test tube 5 and mix the contents. Rinse the graduated cylinder.
5. Transfer 2 mL of the solution from test tube 5 to test tube 4 and mix the contents. Continue in this manner until you have transferred 2 mL of the solution from test tube 3 to test tube 2 and mixed the contents.

CAUTION:
Be sure to rinse your graduated cylinder between each dilution.

6. Number a corresponding set of test tubes from 1 to 6. Using a transfer pipet place 1 mL of each of these standards into a correponding tube.
7. Add 7 mL of 0.1 *M* KOH to each of the six test tubes.
8. Before the addition of base, the test tubes contained the following concentrations. Use these concentrations to plot the standard curve.

Test tube	Concentration
1	0 nmol nitrophenol/mL
2	100 nmol nitrophenol/mL
3	200 nmol nitrophenol/mL
4	400 nmol nitrophenol/mL
5	800 nmol nitrophenol/mL
6	1600 nmol nitrophenol/mL

Part C: Standard Curve and Concentration Measurement

1. Allow the spectrophotometer to warm up at least five minutes prior to taking the readings.
2. Set the wavelength to 410 nm. You will read the standard curve first. Use test tube 1 of the standard as the blank (zero absorbance). Adjust your spectrophotometer as instructed in lab.
3. Read and record the absorbance of each of the nitrophenol standards (test tubes 2 through 6).
4. Read and record the absorbance (*A*) of each time point in your enzyme reaction assay (test tube numbers 1 through 10).

Name: _____

RESULTS

1. Record the absorbance for the standard curve.

Test tube	Concentration	Absorbance
1	0 nmol nitrophenol/mL	
2	100 nmol nitrophenol/mL	
3	200 nmol nitrophenol/mL	
4	400 nmol nitrophenol/mL	
5	800 nmol nitrophenol/mL	
6	1600 nmol nitrophenol/mL	

2. Prepare the standard curve for the assay. Plot the absorbance for each standard tube numbers on the y-axis as a function of the concentration of nitrophenol in the standards on the x-axis.

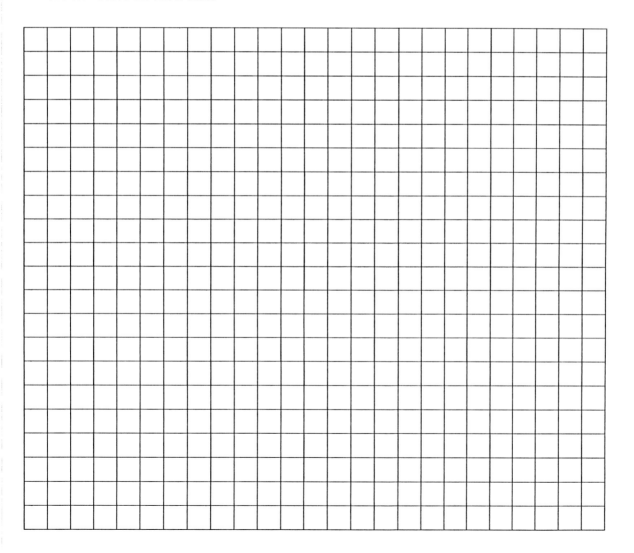

3. Using this standard curve, determine the nanomoles of nitrophenol produced in each tube of your enzyme assay.

Time (min)	Test tube	Absorbance	Nitrophenol produced (nmol)
0	1	0	0
1	2		
2	3		
3	4		
4	5		
5	6		
7.5	7		
10	8		
15	9		
20	10		

4. Plot the nanomoles of nitrophenol (y-axis) produced against time (x-axis).

Name: _____

3. From the above graph, calculate the initial velocity (v_0) of the enzyme reaction in nmoles nitrophenol produced/minute. Draw a tangent to the initial portion of the curve, where it is increasing steeply. The slope of this line is the initial velocity.

$$\text{Rate} = \text{slope} = \frac{\text{change in } y}{\text{change in } x}$$

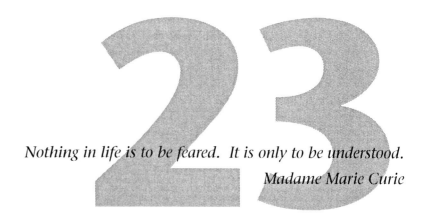

Nothing in life is to be feared. It is only to be understood.
Madame Marie Curie

RADIOACTIVITY and NUCLEAR MEDICINE

In the late 1890s the scientists Antoine Becquerel, Marie Curie, and Pierre Curie discovered radioactivity (Marie Curie coined the word to describe the phenomenon where certain elements spontaneously give off radiation). In 1899 Ernest Rutherford identified two forms of radiation—alpha (α) particles and beta (β) rays; soon after that a third form, gamma (γ) rays, was discovered. These particles are given off during nuclear reactions. Chemical reactions are not to be confused with nuclear reactions. In a chemical reaction the nucleus of the atoms remains unchanged, but in a nuclear reaction a new nucleus and thus a new element is formed. A nuclear reaction is accompanied by the release of radiation in the form of α, β, or γ radiation. The table below shows these three types of radiation, what the nucleus emits in each type, its charge, and its symbol.

Types of Radiation			
Radiation	Components	Charge	Symbol
Alpha	2 protons, 2 neutrons	2+	α
Beta	electron	1-	β or e^-
Gamma	electromagnetic radiation	0	γ

It is important to know type of radiation emitted when using a particular radioisotope. The different types of radiation have different penetrating powers. For example, alpha particles are slow moving and have little penetrating power, so if they are kept in a glass container they are stopped by the glass and are fairly safe. The problem is different if radioactive material emitting alpha particles is ingested. Alpha particles are large and highly charged, so they can cause a lot of damage inside the body where they can get close to organs or important biological molecules. Beta and gamma particles are faster moving and more penetrating. Beta particles are stopped by plastic, so those who work with radioisotopes that emit beta particles keep them behind a plastic shield. Gamma rays have great penetrating power, and the handler must use lead shields for protection.

Radioisotopes are used in medicine for treatment and diagnosis. For example, iodine-131 is used to treat thyroid cancer, technetium-99 is used for diagnostics, and fluorine-18 is incorporated into glucose for PET (positron emission tomography) imaging. Exposure to radiation must be limited. Why? Because the radiation emitted by radioisotopes have the power to do great harm. The particles emitted are very energetic and can collide with a variety of molecules in the body, causing damage to the individual or to their genetic material. Damaged genetic material may pass defects on to future generations. People who work with radioactive compounds must be aware of the type of radiation being emitted so they can protect themselves accordingly.

> Nuclear medicine: a branch of medicine where small amount of radioactive isotopes are used to treat or diagnose disease.

One of the biggest problems associated with using nuclear materials is the nuclear waste they produce. Many radioisotopes used have very long half-lives: the time it takes for half of the original sample to decay. Each radioisotope has a characteristic half-life. That is, if a radioisotope has a half-life of five days, then half of the original sample will decay in five days. In five more days (ten days total) half of the remaining half will decay so that one quarter ($0.5 \times 0.5 = 0.25$) of the original sample remains.

> The half-life of a radioisotope is the amount of time it takes for half of the original sample of the radioisotope to decay.

This lab consists of two parts. In the first part you will explore the concept of the half-life. In the second part of the lab you will look at various everyday articles that contain radioactive materials. You will determine the type of shielding required to stop the radiation. The two parts do not need to be done in order so you might do whichever activity is available, just be careful to put your data in the appropriate section of the data sheet.

MATERIALS

Equipment

100 pennies with container

Geiger counter

Various shields (cardboard, plexiglass, glass, lead, etc.)

Radiation sources

PROCEDURE

Part A: Determination of Half-Life

1. Count the number of penneis in the box and put this number in the data table as Drop 0.

2. Put all the pennies in the box. Shake the box and dump the pennies onto your work surface. Count the number that landed with the head side up and record this number for Drop 1. Place these pennies back in the box. The tails represent the decayed radioisotopes in your sample.

3. Repeat Step 2 with the pennies left in the box. Record the number of pennies that land head side up as Drop 2.

4. Continue until no pennies are left (or eight drops, whichever comes first).

Part B: Testing Radioactive Sources and Shielding

1. Determine the level of background radiation. Turn on the Geiger counter and measure the radiation detected for one minute (counts per minute, cpm). Record this in the Data table. When testing sources and different types of shields you will be subtracting the background level from the radiation detected so that you are only considering the level of radiation due to the source.

2. Place a radiation source 10–12 cm in front of the Geiger counter. Measure the radiation level emitted by the source for one minute and record this level in the Data section.

3. There are several types of shielding available. Place one type of shield half-way between the source and the Geiger counter. Do not move the source because the level of radiation will differ with the distance from the Geiger counter. Again, measure the radiation emitted for one minute and record this in the Data table. Do not forget to note which type of shielding you are using.

4. Repeat this process for at least two other types of shielding.

5. Using a different radiation source, repeat Steps 2 to 4.

Name: _____

DATA

A. Determination of Half-Life

Drop #	Number of pennies left heads up
0	
1	
2	
3	
4	
5	
6	
7	
8	

Using the grid on the next page, plot the Drop # (*x*-axis) *versus* number of pennies left heads up (*y*-axis). Draw a smooth curve through the points.

Part B: Testing Radioactive Sources and Shielding

Background radiation: _____

Testing of Source 1:

Source type 1: _____

Type of shielding	Radiation level (cpm)	Radiation emitted by source (subtract background)
No shield		

Testing of Source 2:

Source type 2: _____

Type of shielding	Radiation level (cpm)	Radiation emitted by source (subtract background)
No shield		

RESULTS

Graph of simulated radioactive decay

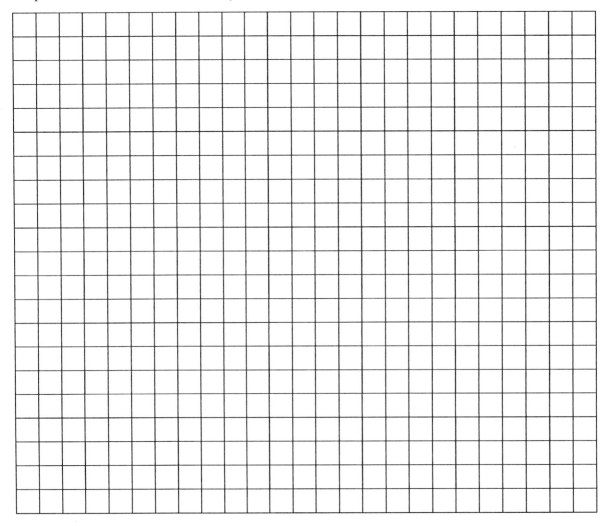

QUESTIONS

1. What is the "half-life" of the pennies; that is, how many drops does it take until you have half (or close to half) as many pennies left as you had when you began?

2. How would the graphs of the radioactive decay of Uranium 238 (half-life of 4.5×10^9 years) and Thorium 234 (half-life of 24 days) differ (if charted on the same graph)? You may either describe the difference or sketch the graph itself.

3. What type of shield was most effective in blocking the radiation from each of your sources?

4. From the results from the shielding can you make any conclusion about the type of radiation emitted by your different source?